D1325610

ON-LINE EVALUATION OF MEAT

CAMPUS LIBR᾿

ırn or renewal o᾿ the la᾿ ᾿ate ᾿᾿
led e᾿᾿ ᾿᾿ ᾿ h᾿ ᾿᾿

HOW TO ORDER THIS BOOK

BY PHONE: 800-233-9936 or 717-291-5609, 8AM–5PM Eastern Time

BY FAX: 717-295-4538

BY MAIL: Order Department
Technomic Publishing Company, Inc.
851 New Holland Avenue, Box 3535
Lancaster, PA 17604, U.S.A.

BY CREDIT CARD: American Express, VISA, MasterCard

PERMISSION TO PHOTOCOPY–POLICY STATEMENT

Authorization to photocopy items for internal or personal use, or the internal or personal use of specific clients, is granted by Technomic Publishing Co., Inc. provided that the base fee of US $3.00 per copy, plus US $ .25 per page is paid directly to Copyright Clearance Center, 222 Rosewood Drive, Danvers, MA 01923, USA. For those organizations that have been granted a photocopy license by CCC, a separate system of payment has been arranged. The fee code for users of the Transactional Reporting Service is 1-56676/95 $5.00 + $ .25.

On-Line Evaluation of Meat

LIBRARY
NATURAL RESOURCES INSTITUTE
CENTRAL AVENUE
CHATHAM MARITIME
CHATHAM
KENT ME4 4TB

H. J. SWATLAND, Ph.D.

Ontario Agricultural College
University of Guelph

664·
907
SWA

TECHNOMIC
PUBLISHING CO., INC.
LANCASTER · BASEL

On-line Evaluation of Meat
a **TECHNOMIC** publication

Published in the Western Hemisphere by
Technomic Publishing Company, Inc.
851 New Holland Avenue, Box 3535
Lancaster, Pennsylvania 17604 U.S.A.

Distributed in the Rest of the World by
Technomic Publishing AG
Missionsstrasse 44
CH-4055 Basel, Switzerland

Copyright © 1995 by Technomic Publishing Company, Inc.
All rights reserved

No part of this publication may be reproduced, stored in a
retrieval system, or transmitted, in any form or by any means,
electronic, mechanical, photocopying, recording, or otherwise,
without the prior written permission of the publisher.

Printed in the United States of America
10 9 8 7 6 5 4 3 2 1

Main entry under title:
  On-line Evaluation of Meat

A Technomic Publishing Company book
Bibliography: p.
Includes index p. 343

Library of Congress Catalog Card No. 95-61607
ISBN No. 1-56676-3339

To My Parents:
for all the Roast Beef, Yorkshire Puddings,
Pheasant, Claret, and Stilton.

CONTENTS

ON-LINE EVALUATION OF meat is a term now used to denote a particular class of operations in the meat industry. The meat may be in any form: intact carcasses on an overhead rail, cuts of meat on a conveyer, comminuted products in a box, or a slurry or batter in a pipe or chopper in the manufacture of processed meat products. But the evaluation must be sufficiently rapid to keep pace with processing line speeds in major plants, so that *on-line* implies both computer interfacing and meat-plant line applications. The evaluation must be objective, or instrument-based, without a human judgement, as well as relatively nondestructive. Above all, the evaluation must relate to a commercially important property of meat: evaluations with an immediate economic value currently are driving the technology, but also we need to collect a folio of long-term research investments for future development. Thus, meat-yield prediction from diode fat-depth probes, TOBEC, (total body electrical conductivity) and VIA (video image analysis) lead the way commercially, but building on this foundation to incorporate a wide variety of meat quality indicators is equally important.

If we refer to methods for yield grading as phase 1 of this technology, now we are starting phase 2 for meat quality. Enthusiasm, coupled with rigorous attention to the principles of scientific methodology and engineering, could take us a long way.

This book aims to provide:

- a review of what has been achieved in phase 1
- a frank exposé of phase 1 problems and possible solutions
- a look at machines and robots for future automation
- an introduction to phase 2 possibilities, including
  - fiber-optics
  - electrical impedance

- lasers
- electromagnetic scanning
- photodiode arrays
- ultrasonics
- video analysis
- electromechanical probes
- off-line fat analysis for boar taint
- economic aspects and commercial incentives
- practical aspects of instrument engineering

The applications of on-line evaluation of meat include almost every aspect of the meat industry:

- carcass yield grading
- quality control
- least-cost formulation
- tenderness and texture
- detection of olfactory taints
- meat color and discoloration
- fat color and softness
- fat content
- marbling fat
- water-holding capacity during storage and processing
- genetic improvement of meat quality
- cooking
- process control sensors

Not only do we hope to improve our meat animals for quality-conscious customers, but soon we will be sorting and streaming the final product for niche markets and maximum economic gain.

Grouped at the end of the book are some plates to show what on-line sensors look like in operation and a series of plates that illustrate components of meat microstructure that are important in understanding the operation of meat probes.

Many thanks to Dr. Joe Eckenrode of Technomic Publishing who encouraged me to write this book. It was assembled during a hectic 10 weeks of an already busy teaching semester, thanks to 89 understanding students who let me cut a few labs and consulting hours. I hope readers can forgive my shaky free-hand sketches, home-made plotting software, and dusty micrographs, plus errors and omissions in the text attributable to my lamentable grasp of real science. Special thanks to Kimberly

Martin and Douglas Bishop for speedy and effective editing and production, and to the eagle eye of my pal Bridget who, at ten paces, can spot mistakes in my writing as effectively as she can see specks of spinach when I attempt to wash her Wedgwood dinner plates.

CHAPTER 1

Grading Carcasses and Cutability

INTRODUCTION

THERE ARE A few essential terms used throughout this book that need an immediate introduction, for readers unfamiliar with the meat industry coming to the subject from another discipline. In the meat industry there are two terms used almost on a daily basis:

- *PSE* stands for pale, soft, and exudative. It is a long-standing problem in pork and a growing concern in some white poultry meats and may occur as a rarity in dark meats such as beef and lamb. PSE meat appears pale because of a high degree of light scattering caused by a low pH; it is soft because of free fluid between the muscle fibers and other factors, and it is exudative, losing weight by drip and evaporation. The costs of increased exudation from PSE meat may be measured in millions of dollars for a major pork packer.
- *DFD* stands for dark, firm, and dry and is the opposite condition to PSE. The meat appears dark because it has a high pH and scatters less light than normal; it is firm because its fluid-filled muscle fibers are still turgid, and it seems dry when eaten. The dryness is misleading, however, because DFD meat has lost less fluid than normal and, thus, may have a higher water content than normal. But the water is held tightly between meat proteins at the microstructural level, resulting in a lack of normal juiciness when the meat is eaten. DFD is a minor problem in pork but is sometimes a major problem in beef, where it may be called dark-cutting beef.

On the scientific front, there is another essential term:

- ATP stands for adenosine triphosphate, an energy-rich molecule providing energy for muscle contraction, the pumping

1

of ions across membranes, and a myriad of other vital functions for a living cell. When the third (tri-) phosphate is added, electrostatic repulsion between phosphorus atoms must be overcome, thus releasing energy when the phosphate is later cleaved by hydrolysis.

The first three terms are universal, without alternatives, but there is another concept in common use, which has a variety of names:

- *PSS*, standing for porcine stress syndrome, is used throughout this book to indicate pigs carrying the mutant HAL 1843 gene that alters calcium release channels in muscle, causing them to undergo uncontrolled glycolysis. The heat and acidity this produces may kill a pig or produce severe PSE after slaughter. Other terms in common use for this condition include stress susceptibility and malignant hyperthermia, while the calcium release channel may be called the ryanodine receptor. Contrary to popular belief, pork from PSS pigs is not always PSE, and all PSE pork does not originate from PSS pigs. For a population that contains numerous homozygous and heterozygous PSS pigs, however, there is no doubt that PSS and PSE are correlated.

Chapter 2 will delve into specifics on PSE and DFD, but more complete explanations of PSE, DFD, and PSS and the role of ATP in muscle contraction and the development of rigor mortis have been provided elsewhere (Swatland, 1994) and are beyond the scope of this book. PSE and DFD may be measured on-line, but they also may be a major source of error when we are trying to measure something else, as will be seen in the latter part of this chapter, which is concerned with measuring the fat content of a carcass. If the loin muscle of a pork carcass is severely PSE, it may be so pale that it is similar in appearance to the back-fat, which brings us to the main topic of the chapter.

A major problem facing the modern meat industry is the difficulty of predicting the yield and quality of meat from an outward inspection of a carcass, although the judgements of an experienced meat buyer may be quite useful. From the degree of skeletal ossification may be estimated the animal's age or maturity at slaughter; from indications of a pizzle eye or shape of the aitch bone, the sex of the animal; and from overall shape, weight, and fatness of the carcass, the type of animal or breed. Combining these clues with experience, a reasonable prediction may be made about the anticipated tenderness, color, marbling, and water-holding capacity of meat – but only reasonable in the realm of averages and probabilities. There is a wide margin of error for any subjective judgement of individual carcasses, and subjective predictions are poorly suited for use in quality control programs because they may change with the mood or fatigue of the individual. For a decision that carries significant economic implications, such as paying the producer of a carcass or choosing meat for a value-added premium treatment, it is preferable that the decision be made with an instrument without any subjective bias. Thus, the ideal system for on-line evaluation of meat yield would produce accurate, nondestructive, real-time estimates and would be inexpensive to install and operate.

Customers at the meat counter prefer meat to be lean, without thick layers of fat between and around muscles. But a variety of quality problems have become apparent as meat has been made progressively leaner. For lean pork and beef, our main concerns are with pH-related quality problems and toughness, respectively. Thus, as the industry is driven towards leaner products, there is a need to improve the objectivity and reliability of carcass evaluation, but we still have far to go.

CURRENT MEAT GRADING SYSTEMS

Principles

Current meat grading systems use mostly subjective decisions, but objective measurements already are being used for back-fat thickness in pork grading. As we move from subjective to objective grading, we need to examine carefully how subjective grading works because it contains both explicit and implicit components. In other words, often, we make hidden decisions, without really stating what we are doing. Implicit yield grading occurs when different classes of carcasses are identified, with the implicit assumption from previous experience that one class will have a higher yield of lean meat than another. This originates from common growth patterns of meat animals and poultry. When they are very young, they have a high proportion of bone and visceral weight so that carcass meat yield is low. As they grow, they add muscle bulk, and the meat yield increases. But if they reach or pass maturity, they are probably involved in breeding or milk or egg production, and muscle bulk may be reduced, giving a low carcass yield. In contrast to this implicit decision making, explicit yield grading is when a particular measurement is made to predict meat yield, usually within one class of animals.

Explicit yield grades are based on a simple principle, established many years ago by Callow (1948), that meat yield is directly proportional to carcass weight but is inversely proportional to carcass fatness (Figure 1.1). If both carcass weight and fatness are taken into account, the inclusion of further data into a prediction equation may not make much difference (Thompson and Atkins, 1980). Subjective estimates of muscularity (the way muscles bulge from the carcass profile) are useful in subjective grading (Perry et al., 1993), but it may be some time before we can achieve this reliably by video image analysis (Chapter 11).

Fat depth now is well established as a basis for predicting meat yield in many countries, and there is a well developed industry in manufacturing and servicing probes, so there is no harm in taking a critical approach to the technology—not to oppose it in any way, but to help improve it. Using Callow's (1948) vintage data, so as not to aim criticism at any particular group at the present time, the most immediate objection is obviously the scatter around the regression line. A simple probe using a regression line to predict meat yield can only achieve maximum reliability on the same population as that on which it was calibrated. Change the population, by age structure, growth attainment, or geography, as easily may happen in practice, and the relationship may become unreliable. The challenge is to develop the technology that

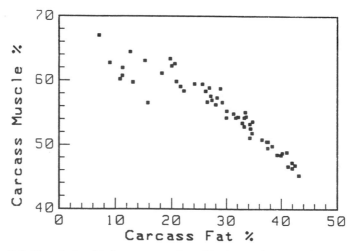

Figure 1.1 The relationship between muscle and fat for a wide range of beef, pork, and mutton carcasses (Callow, 1948).

will allow a system to detect and adjust for the type of population on which it is used.

Bone is another problem. When whole carcasses are dissected into muscle and fat to calibrate a relationship between meat yield and fatness, is bone always a constant proportion of carcass weight? Figure 1.2, again based on Callow (1948), shows a curvilinear relationship of fat to bone that might exist, yet be masked by a narrow range. If the proportion of bone is not constant, this will threaten the reliability of predicting the yield of lean meat from fatness. Butchers may regard bones as piles of inert material to be sold at a loss, but the physiologist looks at bone as a very dynamic reservoir of calcium salts and as a highly vascular tissue with a major involvement in forming new blood cells, particularly in young animals. Bone mass may not be as constant as is sometimes assumed, and the great weight of bone influences hot carcass weight, which is a major component of predicting lean yield.

Another problem in predicting meat yield from fatness is to find a simple, yet reliable, measure of carcass fatness. In subjective grading systems, although fat-depth measurements are made at a constant anatomical site, the grader also is required to adjust subjectively any measurement that does not represent the level of fat on the remainder of the carcass. This is a difficult decision to make subjectively (Murphey et al., 1983) and would require considerable progress in machine vision before it could be done automatically. Complications ensue if some of the carcass fat has been removed before grading, usually by a meat inspector trimming off bruises or tissues with minor infections. Excessive amounts of subcutaneous fat also may be lost to a hide-puller, so that the subjective evaluation of the fat cover on a trimmed carcass may be biased (Murphey et al., 1983). Again, this calls for a subjective response by the meat grader who is scoring fat thickness, and this is difficult to achieve automatically. Most importantly, fat depth is a simple linear measurement, while total carcass fat is a complex anatomical volume, and it would be surprising if the former were a perfect indicator of the latter.

Beef Grading

Although the primary interest in this book is the automated on-line evaluation of meat, subjective grading cannot be ignored. The existing subjective grading systems, such as those administered by the USDA and Agriculture Canada, provide the platform for the newer technology

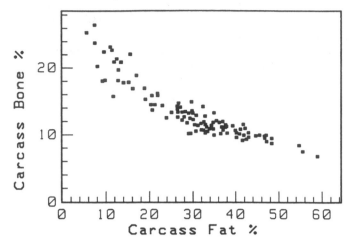

Figure 1.2 *The relationship between bone and fat for a wide range of beef, pork, and mutton carcasses (Callow, 1948).*

and its aspirations. Thus, we need to have a basic understanding of the goals of subjective grading, so that we may project forward in time to set the goals for on-line evaluation of meat. Also, as new systems for on-line evaluation of meat are introduced, compatibility with existing subjective grading systems may be required.

The purpose of subjective grading is to describe the value of a carcass in clearly defined terms that are useful to the meat industry, using federal regulatory agencies as an impartial third party to establish the grades. But if on-line evaluation becomes completely successful, it may be possible to achieve this impartiality from an approved instrument, which will lead to some drastic changes in the general approach to grading. Thus, weights and measures enforcement, which allows buyers and sellers to interact over a weigh-scale of specified accuracy, may provide a role model of how meat grading will be supervised in the future. Even today, if a buyer and seller have established their own system of payment for high- and low-value carcasses, they may dispense with federal carcass grading, and self-enforcement could be even more attractive as reliable on-line instrumentation becomes available.

Another function of existing grading practices is to facilitate long-distance transactions and contracts for future shipments, where the buyer is unable to examine personally the carcasses for sale. With on-line evaluation, this function could be greatly enhanced by a variety of technical options, such as down-loading actual grading data from seller to buyer separated geographically or using new technology to produce carcasses to set specifications for a future contract sale.

In theory, the three major factors used currently to determine the value of a carcass relative to market conditions are carcass weight, meat yield (cutability), and meat quality. All three factors are continuous variables measured in either absolute terms, such as weight, or in relative terms, such as those that might be used by customers. But the number of final grade categories is fairly limited, so the continuous spectrum of carcass properties is subdivided into a relatively small number of grades in a stepwise sequence. Ideally, carcasses placed in the same grade should only exhibit minor differences, while carcasses in different grades should differ in a commercially meaningful manner.

For beef grading in the current USDA grading system, there are three major factors: (1) the sex or type of animal from which the class of carcass originated, (2) the age or maturity of the animal, and (3) the amount of intramuscular or marbling fat within major muscles. The class of carcass (steer, bullock, bull, heifer, or cow) is determined first, identifying male carcasses from the presence of a pizzle eye of a detached penis, irregular rough fat in the cod region, and only a small area of lean exposed ventrally to the pubis or aitch bone. Unlike bullocks and bulls, steers have only a small pizzle eye, with a muscle that is smaller, light red in color, and fine in texture. Bulls have more advanced ossification than bullocks. Heifers are separated from cows by having a smaller pelvic cavity and a more curved aitch bone. Some of these features may be detectable by image analysis, but whether or not accurate decisions can be made by computer remains to be seen.

Next, the beef carcass of an identified class is placed into one of five possible maturity groups, using features of the skeleton and lean meat. In animals about one year of age, the interiors of the vertebral centra are soft, red, and porous in appearance, and the medial surfaces of the ribs are rounded and streaked with red. As cattle grow older, the interiors of their bones become harder, whiter, and less porous. Carcasses from young animals exhibit a maximum of relatively soft cartilage, particularly on the tips of the dorsal spines of the thoracic vertebrae, while older animals acquire hard and ossified cartilage in these locations. In young animals, the sacral vertebrae are only loosely fused, while older animals have their sacral vertebrae solidly fused together. The concentration of myoglobin increases as cattle grow older, and old cattle tend to have dark muscles. But remember that young cattle also may produce carcasses with dark meat (dark-cutters) if they have been severely stressed or exhausted prior to slaughter (Chapter 5).

Subjective color scores for beef are influenced by ambient lighting conditions when the meat is examined (Kropf et al., 1984), which is not a primary problem in on-line color evaluation, where the emission spectrum of the source and the response of the photometer both will be taken into account (Chapter 3). But it may create secondary problems if meat that has been properly sorted by an on-line system is then evaluated subjectively by a purchaser under an inappropriate source of illumination.

In older cattle, the muscle fasciculi forming the grain of the meat are grouped into large and easily detected units to create a coarse subjective texture, while the smaller and more tightly packed fasciculi of younger animals give the lean a firm texture. This type of textural assessment may be undertaken quite readily in the laboratory by image analysis of cut meat surfaces and could be incorporated into an on-line evaluation system if necessary. But careful thought is required before making this investment. Texture probably is important to customers, mainly because they associate coarse texture with the toughness of meat from older animals. Is texture still important if tenderness is guaranteed by other aspects of on-line evaluation? Alternatively, will other aspects of on-line evaluation for tenderness automatically stream the carcasses into fine-texture (tender) and coarse-texture (tough) categories?

Cattle may acquire carotenoid pigments from their feed, and these pigments accumulate in the carcass fat, so that fat from older animals tends to have a yellow or amber color. High pigment levels in feed such as fresh forage may cause the fat to become yellow at an early age. Fat yellowness is relatively simple to assess on-line (Chapter 10).

The current USDA maturity groups are from A to E in order of increasing maturity. Groups A and B include young steers and heifers, while groups C, D, and E include mature dairy cows, old breeding stock, and animals with retarded or overfinished growth. The longissimus dorsi is examined between ribs 12 and 13, where the forequarter and hindquarter are separated. If the texture and color of the lean are acceptable, the amount of marbling fat is currently the main factor that determines the quality grade of the carcass.

Degrees of marbling in a beef carcass are described by a series of subjective terms, related to the percentage area of meat that contains marbling fat: very abundant, 25%; abundant, 21%; moderately abundant, 18%; slightly abundant, 15%; moderate, 11%; modest, 7.5%; small, 2.5%; slight, 1.5%; trace, 0.5%; and practically devoid, 0%. Considerable training is required to use these subjective terms properly, and some degree of spatial interpretation is involved, so that the judgement is not a simple function of the area of fat (McDonald and Chen, 1992). But there seems little point in developing an on-line system to do all this when it is possible to proceed immediately to the same objective by a different route, using an on-line measure of marbling to predict customer response.

However, in the current USDA beef grading system, marbling level is a primary determinant of the grade in the various maturity groups. The grades are: prime, choice, good, standard, commercial, utility, and cutter. The top four grades are maturity groups A and B, ranging from prime at $> 13\%$ down to standard at $< 1\%$ fat area in maturity group A and from $> 15\%$ down to $< 1.5\%$ fat area in maturity group B. The bottom four grades are for maturity divisions C, D, and E. Hidden among these carcasses may be some that still have quite reasonable meat, but which are being downgraded because of low marbling fat area. For the commercial grade, maturity groups C, D, and E require > 2.5, < 7.5, and $> 11\%$ fat area, respectively. For the utility grade, maturity groups C, D, and E require 2.5 to 0.1, 7.5 to 0.5, and 11 to 1.5% fat area, respectively. For the cutter grade, maturity groups C, D, and E accept the lowest requirements of 0, < 0.5, and $< 1.5\%$ fat area. In working with on-line systems, the top grades attract the most attention because there are more of them, they are worth more, and they have a bigger impact than the lower grades. But the lower grades are equally important economically because, among these carcasses, there are a few that are being downgraded for one reason or another, but that still may have quite reasonable meat. If an on-line system can identify these carcasses, allowing maximum use of their commercial potential, then an on-line system will directly increase profitability. Thus, to the scientist and engineer, working with the lower grades of carcasses has a particular interest.

The deposition of marbling fat relative to subcutaneous fat is a complex topic. Older ideas that marbling fat is only deposited once other body compartments such as visceral and subcutaneous depots have been filled are no longer tenable with modern cattle. Because of intense efforts to manipulate fat levels in various compartments by means of genetic selection, feeding programs, and exogenous repartitioning agents, we must anticipate some independence of the fat levels in the various compartments of the body. Thus, with Continental European beef breeds, marbling may decrease while subcutaneous fat is unchanged (Lorenzen et al., 1993). This is an important point to keep in mind for developing algorithms for processing on-line data. Thus, at one point in time or with one type of animal, a correlation of subcutaneous fat depth with marbling may exist, so that a prediction equation using fat depth to predict customer responses to meat juiciness may become established. But the algorithm may fail if applied on another population lacking a correlation of fat depth with marbling.

In the USDA beef grading system, carcasses receive two separate grades, one the quality grade just described and the other for the predicted yield of edible meat. Currently, the yield grade is calculated from several features; subcutaneous fat thickness, visceral fat that remains on the carcass, the cross-sectional area of the longissimus dorsi at the separation of the forequarter and the hindquarter, and the hot carcass weight (which may be replaced by an estimate made from cold carcass weight $\times$ 1.02). The subcutaneous fat (in units of 0.1 inch) is measured over the dorso-lateral edge of the sectioned longissimus dorsi. If the grader considers the fat at this point to be a negatively biased sample of the overall subcutaneous fat, amounts of 0.1 or 0.2 inch may be added to the measured fat depth. Also, the grader estimates the amount of fat (as a percentage of carcass weight) that the meat cutter is likely to remove from regions ventral to the vertebral column and around the pelvis when preparing retail cuts of meat. The cross-sectional area of the longissimus dorsi (in square inches) is measured or estimated subjectively. The determinants of the yield grade then are 2.5 + (2.5 $\times$ adjusted fat thickness) + (0.2 $\times$ percent kidney, pelvic, and heart fat) + (0.0038 $\times$ lb hot carcass weight) $-$ (0.32 $\times$ rib eye area inch^2). The result of the calculation is adjusted downwards to the next whole number, giving integers from 1 to 5. Carcasses with the lowest number yield grade have the highest yield of edible meat. These are calculations that an on-line evaluation system might perform better than human operators; however, fat-depth probes developed for pork grading are of little use for beef because of the irregularity and thinness of beef fat (Kutsky et al., 1982).

Beef grading in Canada is administered by Agriculture Canada. Fat reduction has been officially encouraged since 1972 to utilize the maximum potential for lean growth afforded by European beef breeds with a large mature frame size. But, by 1987, responses by supermarket customers indicated that the tenderness of beef was a concern, and the grading system was altered in 1992 to include a measure of marbling and to make it partly compatible with USDA beef grades (McDonell, 1988; *Canada Gazette*, 1992). Marbling is evaluated only in the top grades: grade A must contain at least traces of marbling, AA has slight marbling, and AAA contains small or greater marbling. All three quality grades are from youthful animals with muscles that are bright red, firm, and fine grained and with fat that is firm and white. The quality grade (A, AA, or AAA) is marked on the four quarters of the carcass within a maple leaf badge. Essentially, any technology that is developed for marbling detection could be applied equally well to both U.S. and Canadian beef grades, which eventually may merge into a single system for the North American Free Trade Area.

The current beef grading system in Canada has only two maturity groups. Youthful carcasses have the cartilaginous caps on their first three thoracic vertebrae no more than half ossified; their first five lumbar vertebrae have cartilage or a red line on the spinous process tip; their spinous processes are red and porous when split; their ribs are red, round, and narrow; and their sternebrae are not yet fused. Mature carcasses have their thoracic caps more than half ossified; no cartilage or red lines are on their lumbar vertebrae; their spinous processes are hard, white, and flinty when split; their ribs are wide, flat, and white; and their sternebrae are ossified.

Like the separation of quality and yield grading in the USDA beef grading system, these operations also are treated separately in Canada. With the current grading system, yield grade A1 has >59% lean, A2 has 54 to 58% lean, and A3 has ≤53%, and the yield estimates are obtained by measuring subcutaneous fat depth and rib-eye length and width. The grader uses a special ruler to measure the fat depth (mm) over the fourth quarter of the longissimus dorsi, using some notches on the ruler. Biologically, however, there is no guarantee that the fat will be spread uniformly all over the carcass (Hopkins et al., 1993), so that single site measurements are not likely to be as valuable as a series of measurements sampling the whole fat depot. There are nine fat classes in the Canadian beef grading system, extending from a 4-mm to a 20-mm fat depth, with a step size of 2 mm.

The next step in the Canadian beef grading operation consists of using the ruler to measure the loin-eye length and width, but this is only an approximate measurement where the dimension is taken as less than a reference box marked on the ruler, within the box (2), or greater than the box (3). These measurements then are used with a look-up table on the ruler to obtain a muscle score. With three muscle lengths and three widths, nine combinations are possible to obtain one of four possible muscle scores, so that some muscle scores may have more than one set of determinants. Again, these operations essentially are a tedious way of reaching an output from a look-up table that an image analysis system could reach faster and more reliably. The muscle score then is used together with the fat class in another look-up table to find the estimated lean yield. With nine fat classes entered against four muscle scores, there are thirty-six possible combinations, but a reduced output of only thirteen possible yield grades, because some yield grades have more than one set of determinants. Finally, the estimated lean yield places the carcass as either A1, A2, or A3, which is marked on the carcass in red ink with a roller. This series of operations is characterized by a considerable amount of redundancy as data are gathered, used for a computation, then lost by recombination of similar carcasses. Essentially, we are looking at a system that was developed using regression analysis, but it lacks the hardware and software at the grading site to compute a prediction. Thus, it is a system pleading for automation.

The lesser grades in Canadian beef grading are more simple than the A grades. Grade B carcasses are all from youthful animals that missed the A grade for one reason or another: B1 for those without any marbling or with less than 4 mm exterior fat, B2 for those with yellow fat, B3 for those with poor muscling, and B4 for dark-cutters. Grades D and E, seldom used, are for mature cattle used for meat processing.

In both USDA and Canadian beef grading, only carcasses that have passed veterinary meat inspection can be graded. This is not likely to be a problem in on-line grading, since it would be possible to detect meat inspection stamps on the carcass by machine vision. But, doubtless, there will be improvements in meat inspection procedures as they too are automated. The point to bear in mind, however, is that compatibility of on-line grading systems with meat inspection systems should be engineered into future developments, not patched on later as an afterthought.

The marking of graded carcasses is an essential part of beef grading. The main point about the grade marking on the carcass is that it must be spread across the carcass (because the carcass soon gets broken into different cuts of meat) to be visible to customers. Somewhere on the whole carcass at least, there must be a code identifying the establishment that graded the carcass. This requirement may become more stringent in the future so that a defective robot or fraudulent marking in a particular plant may be found rapidly and exactly.

At present, carcasses are marked in a preliminary manner immediately after being graded. Then the length of the carcass is marked with a roller brand, rotating to leave a continuous track of the marking. The USDA roller brand uses abbreviations for two of the grades (STNDRD and COMRCL), and the grade appears inside the U.S. federal shield. Canners and cutters are not stamped. The number of the yield grade is indicated inside another shield-shaped stamp. Canadian carcasses have a quality grade (A, AA, or AAA) on each quarter and a yield grade as a roller brand down the length of the carcass. Red ink is used for grade A, blue ink for B, and brown ink for D and E.

Pork Grading

Pork grading, as might be expected, is radically different from beef grading at present. In the United States, barrows and gilts are graded on the basis of their meat quality and yield; however, unless an obvious defect such as oily fat is readily apparent, it is difficult to grade intact sides of pork for meat quality. Thus, at present, there is little or no evaluation of pork quality.

The basic USDA pork yield grades are from grade 1, with back-fat < 1.0 inch and > 60.4% yield, down to grade 4 with back-fat > 1.5 inch and < 54.4% yield. The yield for barrows and gilts is predicted from a combination of back-fat thickness over the last rib and a subjective estimate of muscling on a three-point scale (1, 2, and 3: thin, average, and thick, respectively). The depth of the back-fat and skin is measured at the level of the first rib, the last rib, and the last lumbar vertebra, which acknowledges that fat depth may not be uniform. Sows are graded on the basis of average fat depth.

In Canada, pork grades are used to pay a producer for the amount of sellable meat produced, based on an inverse linear correlation of total back-fat with ham and loin yield (Fredeen and Bowman, 1968). The dorsal spines of the thoracic vertebrae remain on the left side of the carcass when it is split into sides, and the fat depth is measured 7 cm from the midline between ribs 3 and 4 with an optical probe. A look-up table with nine weight-class columns and seven yield-class rows is used to calculate the grade or index from a combination of the back-fat measurement and the warm carcass weight. Exceptions are that ridgelings (cryptorchids) grade at sixty-seven, emaciated carcasses grade at eighty, three index points may be deducted for a badly shaped belly, ten index points may be deducted for abnormal fat color or texture, and tissues removed by a meat inspector because of farm-origin defects reduce the carcass weight. The farm of origin is identified by a shoulder tattoo on the pork carcass, and the producer is paid the numerical product of the reported market price, the grade, and the carcass weight. Thus, the maximum index of 114 pays 114% of the day's base price, while the lowest index of 80 carries a 20% penalty. With a spread like this, it is absolutely essential that grades are fair and objective, and this is a major requirement that motivates developments in on-line meat evaluation.

Veal and Lamb Grading

Veal and lamb grading is a far less developed activity in North America than is beef and pork grading, but other countries such as New Zealand have well-established grading systems for these commodities. The USDA system has five grades for veal carcasses: prime, choice, good, standard, and utility. Carcasses in the higher grades have good muscle development and some evidence of feathering (the deposition of fat within the intercostal muscles). The Canadian system for veal grades is based on muscle color for the well-muscled A grades, with A1 having the lightest color and A4 the darkest red. Instrumentation for color measurement is available to justify the premium for pale veal.

In the United States, lamb carcasses are graded as prime, choice, good, utility, or cull, by fat streaks in the flank, maturity, and quality. A yield grade is calculated from the degree of muscle development in the leg, the amount of kidney and pelvic fat to be trimmed off the carcass, and the subcutaneous fat depth over the rib. The Canadian system for lamb and mutton is a simplified version of the yield grading system used for beef.

Poultry Grading

Poultry grading is basically similar in both the United States and Canada. Chickens are separated into their major classes: male or female rock cornish aged four to five weeks, weighing <0.8 kg; male or female broilers or fryers aged five to eight weeks, weighing 0.8 to 1.8 kg; male or female roasters aged more than nine weeks, weighing >1.8 kg; and capons aged more than nine weeks, weighing >1.8 kg. Turkeys are separated into classes for male or female fryers or roasters aged twelve to sixteen weeks, weighing <4.5 kg; medium or young hens aged eighteen to twenty weeks, weighing 4.5 to 7.5 kg; and heavy or young toms aged 20 to 24 weeks, weighing >7.5 kg. Male or female ducks may be broilers or fryers (less than eight weeks, 1.8−2.8 kg) or roasters (less than sixteen weeks). There is a single class for prime geese: young males or females aged fifteen to twenty weeks, weighing 2.5−6.5 kg.

Identification of poultry with tender meat is a complex subjective valuation based on a number of tactile inputs. Indicative of tenderness, the skin is judged to be soft, pliable, and smooth-textured, and the sternal cartilage is judged to be flexible. Intermediate levels of tenderness resulting from advancing age are identified by increasing skin coarseness and decreased flexibility of the sternal cartilage. Classes with an advanced age include the stag or male chicken under ten months, the yearling hen, tom turkeys under fifteen months, the rooster, the mature hen chicken, mature hen or tom turkeys, mature ducks, and the mature goose. All these have tough meat identified subjectively from skin coarseness and stiff sternal cartilages.

Other features seen during and after slaughter also are indicative of the age of poultry. Young chickens have unwrinkled combs with sharp points, while the comb becomes wrinkled with blunt points in older birds. Young ducks have soft bills, while older ducks have hard bills. Plumage, in general, becomes worn and faded in older birds, unless they have just moulted, while subcutaneous fat becomes dark and lumpy under the main feather tracts in older birds. The pelvic bones become less pliable with age, while the scales become larger, rough, and slightly raised. The oil sac is enlarged and often hardened in older birds. With age, male chickens and turkeys may develop long spurs. In ducks and geese, the cartilaginous rings of the trachea become progressively more stiff.

A number of other subjective features are also involved in poultry grading. Birds are rejected before grading or are trimmed if they are bruised or have pathological features, particularly of the skin and musculature. Older birds sometimes have breast blisters or cysts caused by pressure or irritation from the bird resting on its sternum (McCune and Dellmann, 1968; Mayes, 1980). These defects are obvious to customers but might be difficult to detect automatically.

On-line evaluation of poultry poses some interesting problems. Flocks usually are slaughtered together, so that sorting the various class of poultry need not be necessary. Although, if this were possible automatically, it might radically change the way in which slaughtering could be undertaken. The main challenge is the mixture of different sensory inputs, both visual and tactile, upon which the subjective grading decisions are made. The bilateral symmetry of the sternum is important to customers who, although there is nothing wrong with a crooked sternum in terms of eating properties, generally prefer a symmetrical appearance. If there were a financial incentive, it would not be difficult to automate a symmetry assessment.

FAT-DEPTH PROBES

Terminology

A widely used handheld fat-depth probe, the Hennessy grading probe from New Zealand, is shown in Figure 1.3. As the sharp tip of the probe is pushed into the carcass, the flat plate stays on the surface, enabling the depth of the probe in the carcass to be determined electromechanically. As yet, there has been no general agreement on the terminology relating to probe usage, so readers are alerted to the following terms used throughout this book:

- transect: a cut or line made with a meat probe through the tissues of a carcass
- way-in: describing a transect made during the initial penetration of a tissue, as a probe first penetrates the meat
- way-out: describing a transect made during the withdrawal of a probe from the meat

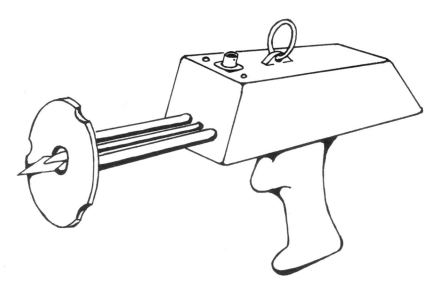

Figure 1.3 Hennessy fat-depth probe from New Zealand.

- signal: the output from a probe sensor, such as a photometer, made during way-in or way-out transects
- depth sensor: a device that measures the depth of the probe in the meat
- depth vector: the signal from the depth sensor during way-in and way-out transects
- vector diagram: the signal from the primary probe sensor, such as a photometer, on the y-axis compared with the depth vector on the x-axis
- cut-off: a point on a vector diagram that represents a change from one condition to another, as when passing from fat to muscle
- threshold: one or more criteria used to locate a cutoff

For example, Figure 1.4 shows the boundary between the back-fat and longissimus dorsi in the thoraco-lumbar region of a pork carcass detected by a typical commercially available diode probe (Destron probe, Anitech, Markham, Ontario) described by Karl (1985) and Usborne et al. (1987). The signal from the photometer diode along a way-out transect is plotted against the depth vector. How it is processed in the Destron probe is proprietary information, but, computationally, we may find the separation between fat and muscle with a simple algorithm, using the depth in the carcass at which the signal passes a cutoff value determined from two signal-dependent scalars, the maximum signal from the fat (F_{max}) and the minimum signal from the muscle (M_{min}), multiplied by an arbitrary threshold of .5

$$\text{cut-off} = ((F_{max} - M_{min}) \times .5) + M_{min}$$

There is considerable variation in signals from conventional diode probes. Reflectance from fat may be irregular if there are traces of muscle or blood or translucent seams of soft fat with a low reflectance, whereas reflectance from muscle may be irregular if it is marbled. The reflectance of PSE muscle may be so high that the boundary between fat and muscle may be indistinguishable.

Limitations

Regrettably, the precise details of the algorithms used to set the cutoff points in commercial fat-depth probes are not public information. This

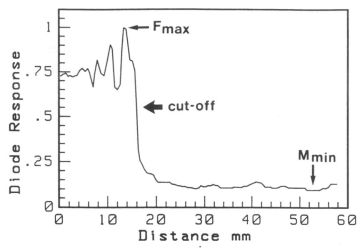

Figure 1.4 *Detection of the cutoff at the boundary between back-fat and longissimus dorsi in a pig carcass, using a position halfway between the maximum reflectance of fat (F_{max}) and the minimum reflectance of lean muscle (M_{min}).*

is one of the fundamental problems we experience in this field, and it deserves a short diatribe.

With the progress of science and technology, scientific instruments have become increasingly sophisticated, and fewer and fewer users have maintained a full understanding of how they work. We may compensate for this by comparing one instrument against another and by measuring known standards, either privately or through publication of results. This is acceptable if the purpose of the instrument is to obtain supporting data, but, if the output of the instrument is the primary consideration of the experiment, there is an inherent danger of the research becoming trivial. It is difficult for a researcher to be critical and innovative or to advance a subject scientifically, if the engineers who produce the instrument know more about the subject than the researcher. At best, the researcher's working hypothesis is reduced to an empirical test of the apparatus, while, at worst, sometimes, nobody really understands what the measurements from an instrument mean.

Where there are commercial opportunities, it is not unknown for an enthusiastic inventor to become involved with a manufacturer and to hit the market with an apparently sound and very useful device for some industrial application. Frequently, this is backed up by a little creative writing from the advertising department, which gives a scientific look to the whole thing. Unfortunately, short-term operations are a lot easier to judge than long-term operations: have your car serviced and you know whether or not it will start before you leave the garage, but have it rust-proofed, and you may never really be sure of the result. Likewise, meat is an extremely difficult commodity to evaluate properly, and it may take considerable effort to test the claims made by the manufacturer of some type of fat-depth probe or meat quality sensor. There are many who take such claims on trust, at least long enough to pay for the probe.

Thus, we may have somewhat of a credibility problem in the on-line evaluation of meat. Some devices may not work as well as claimed, while others work well for specific populations but are of little use on others. A new device that works satisfactorily may produce meaningless predictions after it has been dropped a few times or had a subtle operating feature altered in some way.

From the inventor's perspective, misquotation is a regular ordeal. Administrators or journalists are quite likely, for one reason or another, to report a carefully stated long-term goal as an immediate, all-embracing achievement so that the inventor or researcher is forced to make retractions and denials when faced by potential users of the technology. Let us conclude with the well-known caveat: to avoid disappointment, always read the small print. On-line evaluation of meat has much to contribute to the meat industry, but that is no reason to relax one's critical faculties.

On-line Pork Grading Probes

Fat-depth probes have had a dramatic impact on the economics of pork carcass grading, at least in Denmark and Canada, where they are used to reward producers of lean carcasses, hence providing a powerful incentive for the general improvement of pork carcasses.

Originally two fat-depth measurements were made with a ruler on Canadian pork carcasses, one over the shoulder and one at the posterior end of the loin in split sides. Thus, it was necessary to specify a slight offset to one side in splitting the carcass so that the fat was not lost by irregularity in splitting right from left sides. Two measurements were an insurance against fundamental differences in adipose tissue development between the shoulder and the ham, but they doubled the time for data collection and introduced an extra summation step in computation. Later, the measurement was reduced to a single point for economic reasons. The system was in place to the general advantage of both producers and meat customers and created a steady increase in leanness (Martin et al., 1981).

In the early 1980s, there was still considerable uncertainty as to which direction to proceed, using either invasive or noninvasive electrical probes, optical probes, or ultrasonics. Conductivity probes for detecting the fat-muscle boundary date back to the Lean-Meter of Andrews and Whatley (Hazel and Kline, 1959) in the United States. The Danish KS meter (K for meat and S for fat) detected the fat lean boundary from the difference in conductivity between fat and muscle and was used widely in Danish meat plants (Pedersen and Busk, 1982), but optical probes soon were developed to do the same thing using light reflectance. Essentially, the depth sensor of the conductivity probe was retained, but the boundary sensor was changed from an impedance measurement to a reflectance measurement made between light-emitting and light-sensitive diodes.

The Danish MFA (Meat Fat Automatic) that followed the KS, but that still used the conductivity principle, was tested by Kempster et al. (1979) and found to be competitive with optical probes such as the Ulster probe (developed by the Wolfson Opto-Electronics Unit at Queens' University, Belfast) and the Intrascope (developed for the UK Meat and Livestock Commission). The Ulster probe gave similar results to the Hennessy and Chong fat-depth indicator (FDI) from Aukland, New Zealand (Kempster et al., 1981). To add to the confusion, reasonable results also had been attained with ultrasonics on pork carcasses (Fredeen and Weiss, 1981), but then Jones et al. (1982) and Jones and Haworth (1983) found the Hennessy and Chong FDI of the time to be more precise than the ultrasonic Renco Lean-meter (Minneapolis), and this tipped the balance in favor of optical probes.

Subsequent testing showed the Danish version of the optical probe, the Fat-O-Meater (SFK Ltd, Hvidovre, Denmark), gave similar results to those given by the Hennessy and Chong FDI (Fortin et al., 1984). The maximum prediction of lean meat yield from fat-depth measurements was at a position to one side (6.5 cm) of the midline over the last rib (Kempster and Evans, 1979). The results from noninvasive electronic meat measuring equipment of the day (EMME, Phoenix, Arizona), which previously looked promising for live pigs (Fredeen et al., 1979), were rather poor for pork carcasses (Jones and Haworth, 1983), although still thought to be worthwhile for beef carcasses (Koch and Varnadore, 1976).

Danish Carcass Classification Center

The Danish carcass classification center, first implemented in 1990, is by far the most advanced on-line system for meat evaluation yet developed. The totally automated system (see Plates 2 and 3) evaluates overall carcasses, in addition to separate evaluations of the ham, loin, belly, and forelimb, at line speeds of 360 to 400 carcasses per hour. Carcasses are branded automatically.

When carcasses are moved into the system, their dimensions are measured mechanically (carcass length from gambrel to snout, forelimb position, and height of pubis) to enable correct positioning of a head-holder and the probes. There are nine probes using optical diodes (as shown in Figure 3.8), and the depth vector has a resolution of 0.5 mm in a total length of 18 cm. Probes are inserted by pneumatic pistons: two probes into the ham, two into the belly, two into the shoulder, and three into the loin. All the probes measure fat thickness, while only the loin probes measure meat thickness as well. If a probe hits a bone, the measurement is repeated at a new site 16 mm below the impact site.

As well as being the most advanced system mechanically, the Danish Carcass Classification Center is also the first to exploit neural networks in the analysis of probe signals (Thodberg, 1993). Overall fat depth is taken as a step function between the low reflectance of the probe window outside the carcass and the low reflectance from the muscle below the fat.

Lamb Grading

Jones et al. (1992) used a Hennessy lamb grading probe on 1660 carcasses and concluded that fat-depth measurements between ribs 12 and 13 were accurate enough for on-line grading in Canada. Carcass lean yield was predicted ($R^2 = 0.64$, RSD 28.9 g kg^{-1}), using two fat and two muscle measurements. Kirton et al. (1993) tested several probes and found that they all provided useful predictions of the GR tissue depth measurement over rib 12 at 11 cm from the midline, as used in New Zealand grading. The probes were the Hennessy Grading Probe (Aukland, NZ), the AUS-Meat Sheep Probe (SASTEK, Hamilton, Queensland, Australia), and the Swedish FTC Lamb Probe (Upplands, Väsby, Sweden). The AUS-Meat sheep probe can operate at nine or ten carcasses per minute (Cabassi, 1990).

REFERENCES

Cabassi, P. 1990. *Proc. Aust. Soc. Anim. Prod.*, 18:164.

Callow, E. H. 1948. *J. Agric. Sci., Camb.*, 38:174.

Canada Gazette. 1992. Livestock carcass grading regulations. *Canada Gazette*, Part II, 126(21):3821–3853.

Fortin, A., S. D. M. Jones, and C. R. Haworth. 1984. *Meat Sci.*, 10:131.

Fredeen, H. T. and G. H. Bowman. 1968. *Can. J. Anim. Sci.*, 48:117.

Fredeen, H. T. and G. M. Weiss. 1981. *Can. J. Anim. Sci.*, 61:319.

Fredeen, H. T., A. H. Martin, and A. P. Sather. 1979. *J. Anim. Sci.*, 48:536.

Hazel, L. N. and E. A. Kline. 1959. *J. Anim. Sci.*, 18:815.

Hopkins, D. L., A. A. Brooks, and A. R. Johnston. 1993. *Austral. J. Exp. Agric.*, 33:129.

Jones, S. D. M. and C. R. Haworth. 1982. *J. Anim. Sci.*, 56:418.

Jones, S. D. M. and C. R. Haworth. 1983. *Anim. Prod.*, 37:33.

Jones, S. D. M., O. B. Allen, and C. R. Haworth. 1982. *Can. J. Anim. Sci.*, 62:731.

Jones, S. D. M., L. E. Jeremiah, A. K. W. Tong, W. M. Robertson and L. L. Gibson. 1992. *Can. J. Anim. Sci.*, 72:237.

Karl, W. 1985. *Electron. Products Technol.*, (November):16–17.

Kempster, A. J. and D. G. Evans. 1979. *Anim. Prod.*, 28:87.

Kempster, A. J., D. W. Jones, and A. Cuthbertson. 1979. *Meat Sci.*, 3:109.

Kempster, A. J., J. P. Chadwick, D. W. Jones, and A. Cuthbertson. 1981. *Anim. Prod.*, 33:319.

Kirton, A. H., G. J. K. Mercer, D. M. Duganzich, and A. E. Uljee. 1993. *Proc. N. Z. Soc. Anim. Prod.*, 53:393.

Koch, R. M. and W. L. Varnadore. 1976. *J. Anim. Sci.*, 43:108.

Kropf, D. H., M. E. Dikeman, M. C. Hunt, and H. R. Cross. 1984. *J. Anim. Sci.*, 59:105.

Kutsky, J. A., C. E. Murphey, G. C. Smith, J. W. Savell, D. M. Siffler, and R. N. Terrell. 1982. *J. Anim. Sci.*, 55:565.

Lorenzen, C. L., D. S. Hale, D. B. Griffin, J. W. Savell, K. E. Belk, T. L. Frederick, M. F. Miller, T. H. Montgomery, and G. C. Smith. 1993. *J. Anim. Sci.*, 71:1495.

Martin, A. H., H. T. Fredeen, G. M. Weiss, A. Fortin, and D. Sim. 1981. *Can. J. Anim. Sci.*, 61:299.

Mayes, F. J. 1980. *Br. Poultry Sci.*, 21:497.

McCune, E. L. and H. D. Dellmann. 1968. *Poultry Sci.*, 47:852.

McDonald, T. P and Y. R. Chen. 1992. *Trans. Amer. Soc. Agric. Engin.*, 35:1057.

McDonell, C. 1988. *Ontario Cattlemen's Association, Breeder and Feeder*, 191:14.

Murphey, C. E., H. A. Recio, D. M. Stiffler, G. C. Smith, J. W. Savell, J. W. Wise, and H. R. Cross. 1983. *J. Anim. Sci.*, 57:349.

Pedersen, C. K. and H. Busk. 1982. *Livestock Prod. Sci.*, 9:675.

Perry, D., W. A. McKiernan, and A. P. Yeates. 1993. *Austral. J. Exp. Agric.*, 33:275.

Swatland, H. J. 1994. *Structure and Development of Meat Animals and Poultry.* Lancaster, PA: Technomic Publishing Co., Inc.

Thodberg, H. H. 1993. *Neural Comput. Applic.*, 1:248.

Thompson, J. M. and K. D. Atkins. 1980. *Aust. J. Exp. Agric. Anim. Husb.*, 20:144.

Usborne, W. R., D. Menton, and I. McMillan. 1987. *Can. J. Anim. Sci.*, 67:209.

pH and Meat Quality

INTRODUCTION

JUST ABOUT EVERYONE in meat research and quality control knows that the pH of live muscle, which is usually just above pH 7, decreases after slaughter to values of pH 5.4 to 5.7 in normal meat. The pH of meat has a profound effect on many of its commercial properties: major effects on color and water holding capacity, but only subtle effects on taste and tenderness. Thus, it is difficult to predict beef tenderness using pH (Shackelford et al., 1994b). The pH of meat may be measured on-line with a laboratory gel-filled combination electrode. Rugged portable equipment for routine on-line use in commercial plants is available (NWK Binär GmbH, Landsberg, Germany).

POSTMORTEM GLYCOLYSIS

Glycogenolysis

In living muscle, the complete oxidation of carbohydrate to carbon dioxide and water requires oxygen and releases energy. Much of this energy is captured when ATP is formed from adenosine diphosphate (ADP). Muscle contraction is a primary user of ATP in the living animal, but substantial amounts of ATP are used by the membranes around and within the fiber for maintaining ionic concentration gradients. After slaughter, the body tissues are deprived of oxygen, except in surface layers, and the most they can gain by oxidizing each glucose unit of glycogen is a net gain of two ATP.

Glycogen, the primary storage carbohydrate of muscle, appears in electron micrographs as single granules or clumps of granules located in the sarcoplasm between myofibrils and under the plasma membrane of the fiber. Scanning tunnelling microscopy reveals that glycogen granules are ellipsoidal with a laminar structure; thus, they may grow from one edge, rather than a central point. Glycogen and its associated enzymes are concentrated at the level of the I band.

Although glycogen is a polysaccharide formed from glucose units, the glycogen from some animal tissues is a proteoglucan (glycoprotein) that may contain other monosaccharides and phosphate ester groups. Straight chains of glycogen are formed by carbon $1-4$ linkages, while branch points are formed by $1-6$ linkages. The protein glycogenin acts as a starting block for the formation of new glycogen.

Alternate pathways for glycogenolysis in meat may exist. About 5% may be lysosomal, so that postmortem glycogenolysis may be a combination of both hydrolysis and phosphorolysis, with the former not contributing to the direct formation of lactate.

Phosphorylase initiates glycogenolysis, eroding straight chains of glycogen from their nonreducing ends at carbon 4, and attaching a phosphate group at position 1 as it removes each glucose unit. Phosphorylase erodes straight chains until it comes to the fourth glucose unit preceding a branch point; then the three glucose units before the fourth one that carries the branch are removed together and are added to an adjacent free straight chain so that the $1-6$ linkage, thus exposed at the branch point, may be severed. A second enzyme, debranching enzyme, performs this task. Instead of being released as glucose-1-phosphate, the glucose unit released from a branch point remains as free glucose. Thus, total glycogenolysis liberates glucose-1-phosphate and glucose in a ratio indicating the ratio between the mean length of straight chains and the number of branch points.

Why Lactic Acid Is Formed

After a series of steps in the glycolytic pathway, six-carbon molecules derived from the glucose units of glycogen are split to produce two pyruvates, each with three carbon atoms. All the glycolytic enzymes (except for hexokinase) are concentrated in the I band. If aerobic conditions prevail, pyruvate formed in the cytosol of the muscle fiber enters a mitochondrion and is converted to acetyl-CoA, which then becomes fused to oxaloacetate to form citrate. The citrate then is oxidized in the Krebs cycle, which is completed by the regeneration of oxaloacetate. Continuous activity of the Krebs cycle is fueled by a range of carbohydrates, fatty acids, and amino acids and is the primary system for the aerobic generation of energy. Large numbers of molecules of ATP are produced from ADP by oxidative phosphorylation in the mitochondrial membrane.

In aerobic conditions, the production of two pyruvates from a glucose-1-phosphate results in the reduction of $2NAD^+$. Thus, elsewhere in the muscle fiber, NADH must be reoxidized for glycolysis to continue. Aerobically, this occurs by mitochondrial Krebs cycle activity. But in anaerobic living muscles and in meat, the Krebs cycle is halted, and NADH is reoxidized in the cytosol by lactate dehydrogenase (LDH) during the conversion of pyruvate to lactate. Pyruvate is of no immediate use anaerobically since mitochondrial oxidation has ceased, but its conversion to lactate ensures a continued supply of NAD^+ for the continuation of glycogenolysis and anaerobic glycolysis in the cytosol. However, since these events form only the initial stages of complete carbohydrate oxidation, they do not regenerate much ATP. The net gain of ATP is reduced to two molecules of ATP per glucose-1-phosphate, although glucose released from glycogen branch points generates a total net gain of three molecules of ATP.

Regulation of Glycolysis

The regulation of glycolysis in a muscle fiber of a live animal is integrated with the metabolic state of the fiber and its immediate energy needs. The metabolic state of the fiber is profoundly affected by hormones, particularly epinephrine, and by the extent of recent contractile activity of the fiber. Phosphorylase is particularly important in the conversion of muscle to meat since it may be a primary control site for postmortem glycolysis. Phosphorylase in muscle is most active when it is itself phosphorylated (a form). When dephosphorylated, it is less active (b form). In general terms, therefore, phosphorylase is switched on and off by the addition or removal of its phosphate, with on and off states being relative, rather than absolute. The activity of phosphorylase b is dependent on the presence of AMP, but the activity of phosphorylase a is not.

There are two conflicting requirements that make the mechanism for the activation of phosphorylase rather complex. First, since phosphorylase initiates the release of considerable amounts of chemical energy, there must be safeguards to prevent its uncontrolled activity. In PSS pigs, for example, the uncontrolled activity of anaerobic glycolysis may lead to fatal heat production and acidosis. The conflicting requirement is that the phosphorylase spread through the muscle mass must be rapidly activated by relatively small amounts of epinephrine. The epinephrine activation of severely frightened animals is a "fight or flight" response, and neither of these responses is likely to be of much survival value if the anaerobic energy supply to body muscles is delayed.

The conflicting demands for fail-safe, but rapid, activation are satisfied by two particular features of the activation system. First, the conversion of phosphorylase b to phosphorylase a is inhibited locally in each muscle fiber by high concentrations of ATP and glucose-6-phosphate. Thus, if the energy released by phosphorylase is not rapidly consumed, the energy release system shuts down. If the energy is used, however, AMP and phosphate further enhance the activation of phosphorylase. Second, to enable the rapid activation of phosphorylase throughout the musculature, the relatively small amounts of epinephrine that arrive at the muscle initiate a series of biochemical changes functioning as an amplifier. A small input leads to a large output. Epinephrine causes adenyl cyclase to increase its formation, from ATP, of cyclic AMP. Then, cyclic AMP activates protein kinase. With ATP and magnesium ions present, protein kinase then phosphorylates phosphorylase b kinase b. The active form, phosphorylase b kinase a, in the presence of magnesium ions, finally activates phosphorylase b to phosphorylase a. As a final safety factor, if the supply of inorganic phosphate is inadequate, even phosphorylase a will be relatively inactive, but the system will be primed for rapid energy production once muscle contraction is initiated.

When animals require energy anaerobically during normal activity, phosphorylase b kinase is activated by calcium ions released from the sarcoplasmic reticulum—normally the trigger for muscle contraction in living muscle. Glycogen granules are closely related to the sarcoplasmic reticulum and to glycogenolytic enzymes as part of a structural complex. The activation system linking muscle contraction to glycogenolysis is short-lived to avoid the continuous use and depletion of glycogen reserves. Glycogen is a rapidly available energy source for both brief muscle activity and the early stages of sustained activity. Phosphorylase activity is curtailed by phosphatase, which dephosphorylates phosphorylase a and phosphorylase b kinase a, when muscle activity ceases or an animal recovers from fright. Many features of the system for activating phosphorylase are shared with the activation system for glycogen synthesis, but the shared features are opposite in effect. Thus, the muscle fiber does not attempt to synthesize new glycogen at the same time it is breaking it down, and vice versa.

Porcine Stress Syndrome

PSS pigs have a mutation in the gene for one of the proteins in the calcium-release channel of the sarcoplasmic reticulum (MacLennan and Philips, 1992). The sarcoplasmic reticulum sequesters calcium ions until they are released as the trigger for muscle contraction, so that excessive release leads to excessive contraction, phosphorylase activation, and glycolysis. A test is now commercially available for the detection of the mutation (O'Brien et al., 1993). Leucocytes are isolated by centrifugation and their DNA is freed by digestion of the cells. The polymerase chain reaction is used to produce more than a million copies of the region of DNA around the mutation, which then is cut with a restriction enzyme at the mutation and at a control site for restriction fragment length polymorphism analysis. DNA is separated by electrophoresis and treated with a fluorescent marker to give banding patterns for normal, heterozygous and PSS pigs. The breed frequencies for breeding herds that have been tested in the United States and Canada are given in Table 2.1.

In living animals, the lactate produced by contracting muscles is removed by the circulatory system, but, first, it must travel from the muscle fiber into the intercellular space. Lactate release may be retarded in situations such as exhaustive exercise. A fraction of the lactate leaving a fiber may be in the form of undissociated lactic acid, and this fraction may increase with a rise in external pH. In living animals, blood flow is increased in active muscles (hyperaemia). When hyperaemia occurs as a secondary response to hypoxia, it may be mediated by the release of adenosine into the intercellular fluid between muscle fibers.

Contraction causes an initial drop in muscle fiber pH because of the hydrolysis of ATP. But this may be followed by an increase in pH from

Table 2.1 *Prevalence of PSS in U.S. and Canadian Pig Breeding Herds as Detected by DNA Test (O'Brien et al., 1993).*

Breed	Number Tested	Prevalence of PSS
Pietrain	58	97%
Landrace	1962	35%
Duroc	718	15%
Large White	720	19%
Hampshire	496	14%
Yorkshire	1727	19%
Crossbred	3446	16%

the breakdown of creatine phosphate, which acts as a short-term store of energy with a phosphate that may be transferred to ATP by creatine phosphokinase (CPK).

Overall, now we may see why the pH of meat generally declines after slaughter: the lactate accumulates as a by-product of releasing energy in an attempt to keep the cell alive.

PSE in Poultry

In the author's opinion, PSE may occur in any species capable of decreasing its muscle pH to the isoelectric point while still hot postmortem. Among meat animals, PSE is most noticeable in pork but also may be seen in poultry, particularly turkeys. In pigs, it took many years to sort out the differences between vitamin E-selenium deficiency and PSS and to separate this nutritional myopathy from PSE postmortem. In turkeys, we are still trying to separate PSE (caused solely by glycolysis) from myopathies (with a variety of possible causes).

PSE, with many resemblances to that in pork, has been reported in turkeys (Vanderstoep and Richards, 1974; Van Hoof, 1979; Sosnicki, 1993). In turkeys, there may be a relationship between PSE and the poor slicing color and fragmentation of products such as cooked turkey roll. As considered in Chapter 12, turkeys seem to be deficient in connective tissue towards the end of their phase of rapid muscle growth. If postmortem metabolism is normal and myofibrils retain their fluid, then the weakness of the muscle structure may pass undetected. But if myofibrils release their fluid because the meat is PSE, then the fluid may expand the intercellular space, destroying the weakened adhesion of intramuscular connective tissues and causing product fragmentation after cooking and slicing.

In addition to defects caused by a combination of PSE and inadequate connective tissue development, the muscles of heavy or fast-growing strains of poultry are vulnerable to several myopathies. Deep pectoral myopathy is a direct consequence of successful selection for muscle growth (Siller, 1985). The supracoracoideus muscle becomes trapped by rapid growth within a strong osteofascial compartment between the sternum and the pectoralis muscle. Any vigorous muscle activity causes the muscle to swell, strangulate its blood supply, and become necrotic. Focal myopathy is another disorder of turkey muscle separable from hereditary muscular dystrophy and deep pectoral myopathy (Wilson, 1990). Sosnicki et al. (1988) found oedematous turkey muscles had proliferation of endomysial and perimysial connective tissues, fatty replacement of necrotic muscle, fiber hypercontraction, pyknotic nuclei, and infiltration by mononuclear cells. Wilson et al. (1990) concluded that focal myopathy also was associated with rapid growth. In chickens, right ventricular failure and ascites have also been linked to rapid growth (Julian, 1987), but, here, the cause and effect are easier to visualize.

Thus, at present, we know that PSE caused by low pH may occur in poultry breast muscles, but it is difficult to separate poultry PSE from myopathies because the end result of both mechanisms is a wet, pale muscle lacking a normal texture. Much of the technology used for detecting PSE in pork is applicable to poultry, and all the disadvantages of PSE pork (high fluid loss and poor processing characteristics) are also likely to occur in PSE poultry meat. The best system developed so far for the prediction of PSE in turkey meat is a combination of pH and reflectance, analyzed with a neural network (Santé, 1994).

ON-LINE MEASUREMENT OF pH IN PORK

Historical

PSE pork first got serious attention from the U.S. meat industry in the early 1960s when pH was recognized as a major variable in the quality and, hence, consumption of fresh pork and in the manufacture of processed pork products (Briskey, 1964). Although much of the early research on pork quality showed that normal pigs could produce PSE pork (Bendall, 1973), the study of PSE and DFD became intertwined with studies on PSS. PSS pigs are susceptible to stress, they react positively to the halothane test, and they constantly have high levels of plasma CPK. They were responsible for much of the severe PSE that first became apparent in the 1960s.

As tests for PSS pigs have improved, culminating in the commercial availability of a DNA probe for the mutation in the calcium-release channel of the sarcoplasmic reticulum that causes PSS, the number of PSS pigs has undergone a steady reduction in progressive sectors of the pork industry. At the present time, some pig-producing regions of the United States and Canada have a relatively low incidence of PSS pigs, but the PSE has not gone away. Zhang et al. (1992) concluded that only 20 to 30% of the total variation in meat quality is caused by PSS. Also, it is known that the PSE produced by normal pigs may have an exceptionally high drip loss (Eikelenboom and Nanni Costa, 1988). Thus, the emphasis in the on-line measurement of pH has shifted from trying to establish the seriousness of the PSS problem to attempting to use pH for streaming pork carcasses toward their most efficient utilization. Most of the current in-house research on pH in the meat industry goes unreported, for obvious reasons, so some of the older published pH studies will be used to characterize the general situation.

According to Wismer-Pedersen (1959), as far back as 1914, Herter and Wilsdorf in Germany reported that the increasing wateriness and paleness of pork were causing problems. By 1931, Bech in Denmark had established a meat color evaluation scheme to assess the occurrence of PSE in Danish Landrace pigs, and, by the mid-1950s, PSE had arrived as a scientifically recognized problem (Ludvigsen, 1954; Wismer-Pedersen, 1957).

PSE was thought, at first, to be a pathological muscle degeneration. But then the connection between PSE and rapid glycolysis was discovered: PSE usually develops most severely in muscles that have a pH decline two to four times faster than normal. It was also found that severe PSE may develop in muscles with an abnormally low ultimate pH (pH_u), at pH 4.8 to 5.0, instead of more normal values from 5.4 to 5.7 (Lawrie et al., 1958; Wismer-Pedersen, 1959; Bendall and Lawrie, 1964).

In Denmark, Wismer-Pedersen (1959) proposed that a probe measurement of the pH on the medial surface of the longissimus dorsi of split carcasses might be used to detect potentially PSE carcasses. The earliest measurement feasible on the kill floor was at forty-five minutes postmortem when dressed carcasses entered the meat cooler. A pH measurement on-line at this time has now become almost universally known as the pH_1 value. From the negative correlation of exudate (press-drip) at twenty-four to forty-eight hours postmortem with pH_1, it was proposed that a pH_1 of approximately 6.2 may be used to divide normal from PSE pork. This is appropriate to the majority of the white (fast-contracting, easily fatigued) muscles on the pork carcass, with pH_u values in the range 5.4 to 5.8, but not to the redder muscles, with normal pH_u values > 6.2, which rarely appear to be PSE.

Obviously, the meat industry has undergone some significant improvements in eliminating PSS pigs and increasing line speeds and refrigeration since the days of these early studies. In general, these improvements may have weakened the predictive value of on-line pH measurements (Roseiro et al., 1994). But, in defence of pH measurements, so far they have only been used in a rather simplistic manner.

Thus, the future for on-line measurements of pH may involve superior prediction methods such as neural networks and fuzzy logic, coupled with a host of other on-line measurements of meat quality.

Surveys of pH$_l$

A pH$_1$ < 6.0 generally is regarded as the critical value below which commercially important PSE develops (Taylor, 1966; Bendall et al., 1966). The pH$_1$ index then is defined as the percent of pH$_1$ values below pH 6.0 at forty-five minutes postmortem. In Germany (Schmitten et al., 1983), pH 5.8 has been proposed as the critical value (rather than pH 6.0). This decreases the pH$_1$ index from 72% to 61% when the mean pH is 5.7 but gives nearly identical values of about 15% when the mean pH$_1$ is 6.3. Figure 2.1 shows a histogram for a large collection of pH$_1$ measurements from a plant in Edmonton, Canada (Westra et al., 1979). One of the frustrations of research in this area is that, when the sample size is large enough to be sure that it is representative, very little is known about the animals, because resources seldom permit testing for PSS on such a large scale. Following the logic of pH$_1$ index, the secondary minor peak at pH 5.9 in Figure 2.1 is exactly what we would expect from PSS animals in the general population, but we cannot be sure. Figure 2.2, a remeasurement eighteen hours after slaughter of the same carcasses as

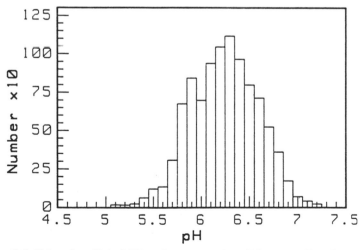

Figure 2.1 *Values for pH$_l$ in 9600 pork carcasses from Edmonton, Canada (Westra et al., 1979).*

in Figure 2.1, shows why pH_1 measurements are superior for the on-line assessment of PSE: the secondary peak of severe PSE carcasses perhaps attributable to PSS pigs has been obscured.

As anyone who has made pH measurements on-line knows, the timing of forty-five or sixty minutes is only an approximation. We start the measuring process by finding a position on-line at which the pork carcasses have been dead for the requisite time and then settle down to take measurements. On a good day, the timing may remain valid, plus or minus a few minutes. But on a bad day, everything from equipment failure to carcasses that have gone adrift in the scalding tank will conspire to exaggerate the plus or minus. Thus, unless the measurements have been made by a team of fanatical time and motion specialists, it may be wise to regard pH_1 and pH_{60} as temporal approximations.

Depending on the point along the line at which the pH measuring station is installed, pH may be measured at thirty, sixty or ninety minutes postmortem (conventionally abbreviated to pH_{30}, pH_{60}, etc.). From Bendall and Swatland (1988), the following equations ($r = 0.90$ and 0.96, respectively) may be used to relate the mean pH_1 to the pH_1 index:

$$\text{for mean } pH_1 > 6.33: pH_1 \text{ index} = 37.8 \times (6.65 - pH_1)$$

$$\text{for mean } pH_1 < 6.33: pH_1 \text{ index} = 110.3 \times (6.39 - pH_1)$$

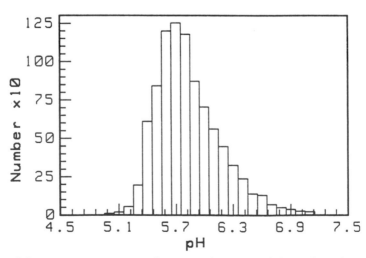

Figure 2.2 *A repeat measurement of carcasses from Figure 2.1 at eighteen hours after slaughter (Westra et al., 1979).*

The first equation may underestimate the required pH_1 index of 50% at $pH_1 = 6.0$ by 7%.

Table 2.2 shows pH_1 values measured on quite large numbers of pigs internationally. From the very extensive published literature on the subject, the reports listed in Table 2.2 were chosen to give a wide geographical coverage and to illustrate typical differences between normal and PSS pigs within breeds. The main point is that, even allowing for differences in measuring technique and time of measurement, there is a tremendous range in pork quality. It is essential to note that the data cannot be used to identify countries or breeds with the highest incidence of PSE. Some of the data with the highest pH_1 indices were taken from countries or breeds that subsequently went on to decimate their PSE levels by remedial action, both in breeding and in slaughtering.

The survey data of Table 2.2 tend to form three categories:

(*1*) Category A with $pH_1 > 6.3$ (mean pH_1 6.43, pH_1 index 8.3): Most of the breeds in category A are British Large White or Large White × Landrace crosses and commercial pigs of Large White extraction. The Canadian Lacombe breed in category A is a cross between Danish Landrace, Berkshire, and Chester White pigs. Category A also includes some Swedish pigs and Chinese native breeds such as Sichuan.

(*2*) Category B with $6.0 > pH_1 < 6.3$ (mean pH_1 6.14, pH_1 index 27.6): Category B contains English, Irish and Danish Landrace pigs, Czech and other Canadian commercial pigs, and Swiss Large White and Landrace pigs (Schwörer et al., 1980).

(*3*) Category C with $pH_1 < 6.0$ (mean $pH_1 = 5.79$, pH_1 index = 66.2): Category C contains English Pietrain and continental Landrace pigs, some of the latter crossed with the Pietrain breed. The main continental Landrace breed appears to come from a different strain than those in Category B.

Obviously, this is far from being a complete survey, and many breeds and countries are unreported. British Yorkshires, Gloucester Old Spot and Hampshires probably belong in category A. For Hampshires, this categorization may be misleading because they show some unusual features, as explained later under the Hampshire effect.

Data such as those in Table 2.2 may appear monumentally boring to someone who is not directly involved in making or analyzing such measurements. But, for someone else who is working on a data set, it becomes compulsively interesting to know what other investigators with

Table 2.2 International Survey of pH_1 and pH_{60} Values Measured On-line in Longissimus Dorsi (LD), Semimembranosus (SM), Gluteus Medius (GM), and Ham Muscles [updated from Bendall and Swatland (1988)].

Place	Year	Type of Pig	Reference[a]	Muscle	pH_1	pH_1 Index	n
England	1964–65	Commercial	Taylor, 1966	LD, SM	6.46	3.9	4737
England	1964–65	Large White	Bendall et al., 1966	LD	6.54	4.0	1360
England	1964–65	Landrace	Bendall et al., 1966	LD	6.42	10.2	952
England	1972–73	Commercial	Kempster and Cuthbertson, 1975	LD	6.55	5.7	6015
England	1973–74	Large White × Landrace	Evans et al., 1978	LD	6.41	7.8	3356
England	1980–81	Commercial	Chadwick and Kempster, 1983	LD	6.38	12.8	5383
England	1973–81	Commercial	Kempster et al., 1984	LD	6.34	12.0	15,487
England	1980–81	Large White	Warriss, 1982	LD	6.62	1.8	171
England	1980–81	Landrace	Warriss, 1982	LD	6.18	32.0	159
England	1980–81	Pietrain	Warriss, 1982	LD	5.67	87.0	42
England	1981–82	Pietrain	Warriss et al., 1983	LD	5.56	80.0	38
England	1964–65	Pietrain	MacDougall and Disney, 1967	LD	5.70	77.0	32
Eire	1977	Large White	Tarrant et al., 1979	LD	6.41	8.0	206
Eire	1977	Landrace	Tarrant et al., 1979	LD	6.26	17.0	151
Eire	1977	Large White	McGloughlin et al., 1979	LD	6.31	10.2	284
Eire	1977	Landrace	McGloughlin et al., 1979	LD	6.14	24.2	258
Denmark	1977–78	Landrace	Nielsen, 1979	LD	6.06	40.2	703

Table 2.2 (continued).

Place	Year	Type of Pig	Reference[a]	Muscle	pH_1	pH_1 Index	n
Denmark	1977–78	Landrace	Nielsen, 1979	SM	6.20	25.0	703
Denmark	1978–79	Landrace	Barton-Gade, 1980	LD	5.98	45.0	2203
Sweden	1978–80	Landrace and others	Lundstrom et al., 1979	LD	5.98	43.0	603
Sweden	1991	Crossbred	Fernandez et al., 1992	LD	6.4	9.4	35
Sweden	1991	Commercial	Karlsson, 1992	LD	5.7	76.1	150
Switzerland	1976–80	Large White	Schwörer, 1982	LD	6.03	40.0	2909
Switzerland	1976–80	Landrace	Schwörer, 1982	LD	6.17	26.0	398
Switzerland	1976–80	Landrace	Schwörer, 1982	LD	5.85	60.0	1736
Germany	1977	Landrace	Scheper, 1977	LD	6.16	16.5	2319
Germany	1977–78	Landrace	Blendl and Puff, 1978	LD	5.67	77.0	832
Germany	1979–80	Landrace	Kallweit, 1981	LD	5.90	53.0	871
Germany	1983	Mixed breeds	Schmitten et al., 1983	LD	5.71	71.0	422
Germany	1988	German Landrace	Li et al., 1989	LD	5.9	54.0	470
Germany	1988	German Large White	Li et al., 1989	LD	6.5	5.7	279
Germany	1988	Pietrain and Belgian Landrace	Li et al., 1989	LD	5.7	76.1	350
Yugoslavia	1977	Mixed breeds	Manojlovic and Rahelic, 1978	SM	6.36	8.5	8229

(continued)

35

Table 2.2 (continued).

Place	Year	Type of Pig	Reference[a]	Muscle	pH₁	pH₁ Index	n
Czechoslovakia	1977	Commercial	Mojto et al., 1978	SM -	6.16	26.0	6647
Canada	1978–80	Commercial	Martin et al., 1981	SM	6.53	3.5	3114
Canada	1978–80	Lacombe	Sather et al., 1981	SM	6.39	9.0	1726
Canada	1978	Commercial	Westra et al., 1979	LD	6.27	21.0	9600
South Africa	1970–71	Large White Cross breds	Naude and Klingbiel, 1973	LD	6.48	6.3	1599
Netherlands	1974	PSS Landrace	Eikelenboom and Minkema, 1974	LD	5.86	58.5	15
Netherlands	1974	Landrace	Eikelenboom and Minkema, 1974	LD	6.34	11.7	90
Netherlands	1974	PSS Landrace	Eikelenboom and Minkema, 1974	LD	5.88	56.2	14
Netherlands	1974	Landrace	Eikelenboom and Minkema, 1974	LD	6.40	9.4	111
Netherlands	1988	Landrace	Eikelenboom and Nanni Costa, 1988	GM	6.28	12.1	70
Netherlands	1988	PSS Landrace	Eikelenboom and Nanni Costa, 1988	GM	5.74	71.7	70
Italy	1987	Large White × Landrace	Russo et al., 1987	LD	6.27	13.2	915
Italy	1987	Large White × Landrace	Russo et al., 1987	LD	6.09	33.1	999
Italy	1987	Large White × Landrace	Santoro and Lo Fiego, 1987	Ham	6.47	6.8	18

36

Table 2.2 (continued).

Place	Year	Type of Pig	Reference[a]	Muscle	pH$_1$	pH$_1$ Index	n
Italy	1987	PSS Large White × Landrace	Santoro and Lo Fiego, 1987	Ham	5.97	46.3	17
France	1985	Large White	Monin and Sellier, 1985	LD	6.39	9.8	15
France	1985	Pietrain	Monin and Sellier, 1985	LD	6.08	34.2	11
France	1985	PSS Pietrain	Monin and Sellier, 1985	LD	5.56	91.5	13
France	1985	Hampshire	Monin and Sellier, 1985	LD	6.36	10.9	17
Spain	1992	Belgian Landrace	Oliver et al., 1993	LD	5.63	83.8	40
Spain	1992	Landrace	Oliver et al., 1993	LD	5.70	76.1	58
China	1987	Sichuan	Xuewei et al., 1992	LD	6.43	7.1	28
China	1987	Crossbred	Xuewei et al., 1992	LD	6.27	28.7	122
China	1987	Duroc, Hampshire and Landrace	Xuewei et al., 1992	LD	6.26	33.3	12

[a]Original publications should be consulted for exact details because data may have been rounded or estimated to simplify the table presented here.

the same objective have found. Which is probably why not many commercial data sets get published for competitors to scrutinize. Unfortunately, public data on the pH of U.S. pork carcasses are few and far between.

Has PSE increased as pork carcasses have become leaner? Evidence from pH_1 surveys of British commercial pigs since 1964 shows a considerable increase in the mean pH_1 index from 3.8% to about 12% over the last twenty years, while there was not much change in breed composition of commercial herds during this period, mostly Large White and Landrace × Large White crosses (Kempster et al., 1984). The increase is unlikely to have originated from stunning methods, which, if anything, improved over this period. Kempster et al. (1984), however, proposed that the increase in PSE might be caused by an increase in the size of abattoirs and, thus, an intensification of the features in the slaughter process that cause PSE to develop, most likely transport and preslaughter lairage. There is little or no evidence for an increase in PSS during this period, although more British Landrace pigs were crossbred with Large White pigs than formerly, and crossbreds had a greater tendency towards PSS than the pure Large White breed.

The three Canadian surveys in Table 2.2 are all from Alberta, with two in category A and one in category B. In the mid-1970s, Ontario would have been in category B, with some major plants having CO_2 stunning. By the early 1980s, however, Ontario plants using manual electrical stunning were well into category A with mean $pH_1 = 6.5$ (Swatland, 1982). From carbon dioxide stunning in the mid-1970s to automated electrical stunning a decade later, Ontario showed a decline in the pH_1 index from 15.8% to 9.0%. Changes to slaughter methods, rather than a reduction in PSS, in Ontario were probably responsible for these changes in pH_1, but it cannot be proven.

For Norway, Frøystein (1981) noted an increase in pig mortality during transport, coupled with an incidence of PSE in a mainly Landrace population. The incidence of PSE was about 19% of all the pigs slaughtered, with variations between factories from 14% to 34%. Norway had only a relatively low (5.4%) incidence of PSS pigs at this time. In Hungary, Hamori (1980) tested 30,880 carcasses on-line and found Hungarian Large White with 10.5% PSE at pork weight and 27.9% PSE at bacon weight, Landrace crossbreds with 29.7% to 49% PSE, Swedish Landrace with 34% to 46% PSE, and Pietrain with 89% to 90% PSE. In Portugal (Santos et al., 1994), relatively large numbers of PSS pigs, combined with hot summer weather, led to high levels of both PSE (30%) and DFD (10%).

Germany has a great tradition of on-line measurement of pork quality. Schepers (1982) found the mean pH_1 of longissimus dorsi decreased from 6.67 in 1958 to 6.30 in 1975 and to between 5.56 and 5.85 in 1980. This indicates an overall rise in the pH_1 index of about $1-2\%$ in 1958 to $58-80\%$ in 1980. But during this time, there was a change in breed composition of the Rhineland herds. The local Landrace breed was dominant around 1958 but was replaced by Belgian Landrace crossbreds and Pietrain purebreds, with a reduction in the incidence of pure German Landrace from 46.4% in 1973 to 3.6% in 1980. By 1980, Kallweit (1981) also reported from North Germany a significant decrease in pH_1 as German Large White (Edelschwein) and Landrace pigs were crossed with Pietrain and Belgian Landrace pigs. The mean pH_1 was 5.9, and 31% of the values were below pH 5.6. Since 1960, the incidence of deaths from various disorders, including transport losses, has increased from 3% to 6% in 1979.

In Switzerland, Denmark, and the Netherlands, the incidence of PSS has been substantially reduced by genetic selection using the halothane test. In Swiss breeding herds since 1976, the incidence of PSE in the Landrace breed has been reduced from 33% in 1978 to less than 10% in 1982, without significantly lowering the proportion of prime cuts on the carcass (Schwörer, 1982: Schwörer and Blum, 1983). In the Netherlands, where the halothane test devised by Eikelenboom first was used on a large scale, the incidence of halothane positive Dutch Landrace pigs was reduced in elite herds from 36.1% in 1977 to 15.2% in 1980 and to 5.2% in 1983 (Eikelenboom, 1984).

In Denmark, from the mid-1970s, meat quality has steadily been improved as PSS Danish Landrace were crossed with halothane-negative English Large Whites, Hampshires, and Yorkshires and with the French Duroc breed. The incidence of PSE decreased from 29.5% to 14.5% between 1975 and 1981 (Staun and Jensen, 1982). The Danish data in Table 2.2 should definitely be considered as out of date, since they describe the largely Landrace populations of the early 1970s.

Hampshire Effect

Both the rate and the extent of glycolysis are important in the determination of pork quality. Thus, both a relatively low pH_1 and/or pH_u may induce PSE. It follows that the reliability of pH_1 as a predictor of PSE depends upon the general existence of a relationship between the rate and extent of glycolysis. On-line measurements of pH_1 detect carcasses that are already well on their way to becoming PSE because of rapid postmortem glycolysis, but pH_1 fails to detect future PSE in carcasses with a normal initial rate but an extended duration of glycolysis (caused by high initial levels of glycogen). Monin and Sellier (1985) and Sellier (1988) refer to this condition as the Hampshire effect. Halothane-negative Hampshires, with a relatively high pH_1 around 6.35, subsequently produce PSE meat with a relatively low pH_u around 5.45. Evidence of late-developing PSE (more than forty-five minutes postmortem) may be found in the statistical data of many published studies and cannot be attributed solely to Hampshire pigs, even though it may be prevalent and determined genetically as a dominant trait in that breed (Sellier, 1988). As well as spoiling pH_1 as an early on-line predictor of PSE, late-developing PSE or the Hampshire effect probably also weakens the reliability of other on-line optical and electronic methods for PSE prediction at forty-five minutes postmortem.

Seasonal Effect

Another point to bear in mind when considering on-line pH data is the possibility of a seasonal effect. It is common knowledge in the meat industry that the PSE problem tends to be worse in hot weather, and this shows in both pH_1 and pH_u (Figure 2.3). In addition to the weather directly affecting an animal's physiology and the refrigeration efficiency of the plant, there are a myriad of indirect seasonal effects such as feed quality and animal age that should not be forgotten. At present, we can only prove a seasonal effect for on-line pH, but we should anticipate seasonal effects in any type of on-line measurement in the meat industry.

DFD AND DARK-CUTTING BEEF

DFD may be a minor problem in pork, but DFD in beef often is a major problem. Dark-cutting beef may be caused by preslaughter stress, transport exhaustion, hunger, fear, climatic stress, or aggressive behavior, particularly between young males. Essentially, any factor that leads to the depletion of muscle glycogen will limit the amount of lactate formed postmortem, causing DFD meat with a high ultimate pH (typically above pH 5.9).

In beef, climatic stress may create a seasonal pattern to the incidence of dark-cutting. Young bulls have the worst reputation for dark-cutting,

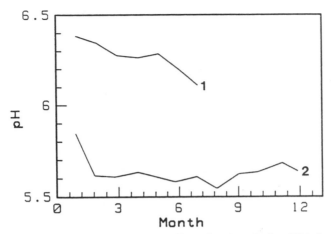

Figure 2.3 *Seasonal effects from January to December (months 1 to 12) in longissimus dorsi pH_1 for Canadian pigs (line 1; Westra et al., 1979) and pH_u for Irish pigs (line 2; Gallwey and Tarrant, 1979).*

particularly in longissimus dorsi, semitendinosus, semimembranosus, adductor, and gluteus medius muscles, whereas, in other muscles, pH values may be near to normal (Tarrant and Sherington, 1980). The effects of preslaughter stress are quite variable, and the heritability of dark-cutting in beef is low (Shackelford et al., 1994a). Although dark-cutting beef is discounted severely in grading because of its visual appearance (Chapter 1), taste panels may report that eye-of-round roasts from dark-cutting beef carcasses are similar to normal beef for softness, juiciness, tenderness, and connective tissue content and that bland taste is its only defect (Hawrysh et al., 1988). When dark-cutting beef occurs naturally, the sporadic pattern of elevated pH between muscles differs from the overall elevation of pH throughout the whole musculature when dark-cutting is induced experimentally by epinephrine injection. Glycogen depletion in young bulls is most severe in fast-contracting muscle fibers, whereas glycogen depletion in slow-contracting fibers is less rapid but persists for longer. Glycogen levels may take several days to return to normal.

Dark-cutting beef may be detected on-line with a fiber-optic probe (MacDougall and Jones, 1981; Gariépy et al., 1994), but early detection may be difficult or impossible because dark-cutting essentially is a failure to develop a normal amount of pH-related light scattering. Thus, all beef looks dark-cutting immediately after slaughter. However, a method that may be used soon after slaughter is to stimulate beef muscle with platinized electrodes (to avoid discoloration) and then to measure pH, with pH > 6.1 indicating dark-cutting (Hald, 1993).

TECHNICAL ASPECTS OF ON-LINE pH MEASUREMENT

There are many potential sources of error in the on-line measurement of meat:

(*1*) Dirty probe electrodes

(*2*) Poor standardization of the electrode against known buffer solutions at regular intervals and poor control of instrument drifting

(*3*) Failure to repeat dubious readings

(*4*) A subjective bias to read lower pH values in meat that is obviously PSE and vice versa for DFD meat

(*5*) Failure to correct for meat temperature

(*6*) Measurement of meat pH from samples with temperature differences of more than 5°C

(*7*) High variance of measurements made on dry muscles with a high pH

(*8*) Failure to check the repeatability of random batches of measurements

(*9*) A false assumption that all pH meters and types of pH electrodes standardized according to the manufacturers' instructions are equally accurate, regardless of battery voltage, operating temperature, internal condensation on circuit components, static fields between a hanging carcass and an operator, and development of high-resistance corroded terminals (Korkeala et al., 1986)

(*10*) Failure to check on-line pH meters against a proven laboratory meter

(*11*) Failure to measure at sufficient depth within the meat to be sure that carcass washing or spray cooling has not cooled the muscle or altered its pH

(*12*) Obstruction or failure of the KCl reference electrode

(*13*) Failure to adopt a standard method to deal with instrument warm-up time, drift, and equilibration time

(*14*) Inadequate washing of the electrode between carcasses

The effect of temperature on pH buffers in meat may be difficult to correct (Bendall and Wismer-Pedersen, 1962). Temperature variation may only be from 37−40°C, causing only small variations in the measured pH of less than +0.02 units, but it may create systematic errors as follows:

(*1*) In carcasses with exceptionally fast glycolysis, because glycolysis is exothermic

(*2*) When the time from the start of slaughter to the pH measurement deviates from forty-five minutes

(*3*) When scalded carcasses are compared with skinned carcasses

(*4*) When measurements made at the start of a day (when refrigeration systems are often at peak performance) are compared with measurements made at the end of the day

(*5*) When some measurements are made deep in a muscle, while others are made near the surface of a muscle recently washed with cold water

Measurements of pH on homogenized meat samples in potassium chloride and iodoacetate solution to stop glycolysis (Bendall, 1973) have a higher repeatability than probe measurements on intact muscles and provide the best means to verify on-line probe measurements made directly from the meat. However, the sample temperature at the time of measurement is usually 18 to 20°C, and this has the effect of raising the pH_i 0.2 to 0.3 units above the values taken with a probe in a warm muscle on the carcass at 37−40°C (Bendall and Wismer-Pedersen, 1962), but only if a temperature correction is made in the probe measurement. For example, in Table 2.2, this may bias the high pH_i values of normal pigs more than it does the low pH_i values from PSS pigs. The same factor may also apply to pH_u measured on a carcass at 1 to 3°C. Any pH_u values taken at 20°C may be about 0.2 to 0.3 pH units lower than the values taken by probe at 1 to 3° if a temperature correction is made with the probe measurement.

The estimated effect of temperature is to lower the pH by 0.1−0.15 units per 10°C rise in temperature and to raise it by the same amount per 10°C fall in temperature (Bendall and Wismer-Pedersen, 1962). This effect is caused by changes in the pK values of carnosine, anserine, and histidine buffers in meat, as the temperature is raised or lowered (Martin and Edsall, 1960; Lentz and Martell, 1964). Hydrolysis of creatine phosphate caused by homogenization causes only small changes in pH (Spriet et al., 1986).

REFERENCES

Barton-Gade, P. A. 1980. *Proc. 26th Eur. Meeting Meat Res. Workers, Colorado Springs,* p. 50.

Bech, N. 1931. *Beretn. Forsøgslab.,* 139:1.

Bendall, J. R. 1973. In *Structure and Function of Muscle,* Vol. II. G. H. Bourne, ed., 2nd edition, part 2, New York: Academic Press, pp. 243–309.

Bendall, J. R. and R. A. Lawrie. 1964. *Anim. Breed. Abst.,* 32:1.

Bendall, J. R. and H. J. Swatland. 1988. *Meat Sci.,* 24:85.

Bendall, J. R. and J. Wismer-Pedersen. 1962. *J. Food Sci.,* 27:144.

Bendall, J. R., A. Cuthbertson, and D. P. Gatherum. 1966. *J. Food Techol.,* 1:201.

Blendl, H. M. and H. Puff. 1978. *Fleischwirtschaft,* 58:1702.

Briskey, E. J. 1964. *Adv. Food Res.,* 13:89.

Chadwick, J. P. and A. J. Kempster. 1983. *Meat Sci.,* 9:101.

Eikelenboom, G. 1984. *64th Annual Conference of the Canadian Meat Council, Quebec,* pp. 1–11.

Eikelenboom, G. and D. Minkema. 1974. *Tijdschr. Diergeneesk.,* 99:421.

Eikelenboom, G. and L. Nanni Costa. 1988. *Meat Sci.,* 23:9.

Evans, D. G., A. J. Kempster, and D. E. Steane. 1978. *Livest. Prod. Sci.,* 5:265.

Fernandez, X., M. Mågård, and E. Tornberg. 1992. *Meat Sci.,* 32:81.

Frøystein, T. 1981. In *Porcine Stress and Meat Quality.* T. Frøystein, E. Slinde, and N. Standal, eds., Å s, Norway: Agricultural Food Research Society, pp. 75–89.

Gallwey, W. J. and P. V. Tarrant. 1979. *Acta Agric. Scand. Suppl.,* 21:32.

Gariépy, C., S. D. M. Jones, A. K. W. Tong, and N. Rodrigue. 1994. *Food Res. Internat.,* 27:1.

Hald, T. L. 1993. *Fleischwirtschaft,* 73(2):140.

Hamori, D. 1980. *Magy. Allatorv. Lap.,* 35:525.

Hawrysh, Z. J., P. J. Shand, and M. K. Erin. 1988. "University of Alberta 67th Annual Feeder's Day Report," pp. 64–66.

Julian, R. J. 1987. *Highlights of Agricultural Research in Ontario,* 10:27.

Kallweit, F. 1981. In *Porcine Stress and Meat Quality.* T. Frøystein, E. Slinde, and N. Standal, eds., Å s, Norway: Agricultural Food Research Society, pp. 90–96.

Karlsson, A. 1992. *Meat Sci.,* 31:423.

Kempster, A. J. and A. Cuthbertson. 1975. *J. Food Techol.,* 10:73.

Kempster, A. J., D. G. Evans, and J. P. Chadwick. 1984. *Anim. Prod.,* 39:455.

Korkeala, H., O. Maki-Petays, T. Alanko and O. Sorvettula. 1986. *Meat Sci.,* 18:121.

Lawrie, R. A., D. P. Gatherum, and H. P. Hale. 1958. *Nature, Lond.,* 182:807.

Lentz, G. R. and A. E. Martell. 1964. *Biochemistry,* 3:750.

Li, X. W. von, K. U. Götz, and P. Glodek. 1989. *Züchtungskunde,* 61:370.

Ludvigsen, J. 1954. *Beretn. Forsøgslab.,* 272:1.

Lundstrom, K., H. Nilsson, and B. Malmfors. 1979. *Acta Agric. Scand., Suppl.,* 21:71.

MacDougall, D. B. and J. G. Disney. 1967. *J. Food Technol.,* 2:285.

MacDougall, D. B. and S. J. Jones. 1981. Translucency and color defects of dark-cutting meat and their detection, in *The Problem of Dark-cutting Beef,* D. E. Hood and P. V. Tarrant, eds., The Hague: Martinus Nijhoff Publishers.

MacLennan, D. H. and M. S. Phillips. 1992. *Science,* 256:789.

Malmfors, G. 1981. In *Porcine Stress and Meat Quality.* T. Frøystein, E. Slinde, and N. Standal, eds., Ås, Norway: Agricultural Food Research Society, pp. 179–184.

Manojlovic, D. and S. Rahelic. 1978. *Proceedings of the 24th European Meeting Meat Research Workers Conference,* A 3:1.

Martin, R. B. and J. T. Edsall. 1960. *J. Am. Chem. Soc.,* 82:1107.

Martin, A. H., H. Fredeen, P. T. L'Hirondelle, A. C. Murray, and G. M. Weiss. 1981. *Can. J. Anim. Sci.,* 61:289.

McGloughlin, P., C. P. Ahern, and J. V. McLoughlin. 1979. *Acta Agric. Scand., Suppl.,* 21:427.

Mojto, J., J. Jedlicka, and S. Palenik. 1978. *Fleischwirtschaft,* 58:1709.

Monin, G. and P. Sellier. 1985. *Meat Sci.,* 13:49.

Naude, R. T. and J. F. G. Klingbiel. 1973. *S. Afr. J. Agric. Sci.,* 3:183.

Nielsen, N. J. 1979. *Acta Agric. Scand., Suppl.* 21:91.

O'Brien, P. J., H. Shen, C. R. Cory, and X. Zhang. 1993. *J. Amer. Vet. Med. Assoc.,* 203:842.

Oliver, M. A., M. Gispert, and A. Diestre. 1993. *Meat Sci.,* 35:105.

Roseiro, L. C., C. Santos and R. S. Melo. 1994. *Meat Sci.,* 38:353.

Russo, V., P. Bosi, and L. Nanni Costa. 1987. In *Evaluation and Control of Meat Quality in Pigs.* P. V. Tarrant, G. Eikelenboom, and G. Monin, eds. Dordrecht, the Netherlands: Martinus Nijhof Publishers, pp. 211–224.

Santé, V. 1994. *Meat Focus Internat.,* 3:403.

Santoro, P. and D. P. Lo Fiego. 1987. In *Evaluation and Control of Meat Quality in Pigs.* P. V. Tarrant, G. Eikelenboom, and G. Monin, eds., Dordrecht, the Netherlands: Martinus Nijhof Publishers, pp. 429–436

Santos, C., L. C. Roseiro, H. Gonçalves, and R. S. Melo. 1994. *Meat Sci.,* 38:279.

Sather, A. P., A. H. Martin, and H. T. Fredeen. 1981. In *Porcine Stress and Meat Quality.* T. Frøystein, E. Slinde, and N. Standal, eds., Ås, Norway: Agricultural Food Research Society, pp. 274–82.

Scheper, J. 1977. *Fleischwirtschaft,* 57:1489.

Schepers, K-H. 1982. 35. Hochschultagung der Landw. Facultät der Univ. Bonn, Bonn.

Schmitten, F., K-H. Schepers, A. Festerling, B. Hubbers and U. Reul. 1983. *Schweinzücht Schweinemast,* 12:418.

Schwörer, D. 1982. Doctor's Thesis, ETH. 6978, Tech. Hochschule, Zurich, pp. 1–138.

Schwörer, D. and J. K. Blum. 1983. *Kleinviehzüchter,* 31:278.

Sellier, P. 1988. Meat quality in pig breeds and in cross breeding, *International Meeting on Pig Carcass Meat Quality,* Reggio Emilia, Italy.

Shackelford, S. D., M. Koohmaraie, T. L. Wheeler, L. V. Cundiff, and M.E. Dikeman. 1994a. *J. Anim. Sci.,* 72:337.

Shackelford, S. D., M. Koohmaraie, and J. W. Savell. 1994b. *Meat Sci.,* 37:195.

Siller, W. G. 1985. *Poultry Sci.,* 64:1591.

Sosnicki, A. A. 1993. *Meat Focus Internat.,* 2:75.

Sosnicki, A., R. G. Cassens, D. R. McIntyre, and R. G. Vimini. 1988. *Avian Pathol.*, 17:775.

Spriet, L. L., K. Sonderlund, J. A. Thomson, and E. Hultman. 1986. *J. Appl. Physiol.*, 61:1949.

Staun, H. and P. Jensen, P. 1982. *Tierzuchter*, 34:496.

Swatland, H. J. 1982. *Can. J. Anim. Sci.*, 62:725.

Tarrant, P. V. and J. Sherington. 1980. *Meat Sci.*, 4:287.

Tarrant, P. V., W. J. Gallwey, and P. McGloughlin. 1979. *Ir. J. Agric. Res.*, 18:167.

Taylor, A. McM. 1966. *J. Food Techol.*, 1:193.

Van Hoof, J. 1979. *Veter. Quart.*, 1:29.

Vanderstoep, J. and J. F. Richards. 1974. *Can. Inst. Food Sci. Technol. J.*, 7:120.

Warriss, P. D. 1982. *J. Food Techol.*, 17:573.

Warriss, P. D., S. C. Kestin, and J. M. Robinson. 1983. *Meat Sci.*, 9:271.

Westra, R., J. R. Thompson, R. T. Hardin, and R. T. Christopherson. 1979. *Proceedings of the 58th Annual Feeder's Day*, University of Alberta, pp. 4–6.

Wilson, B. W. 1990. *Proc. Soc. Exp. Biol. Med.*, 194:87.

Wilson, B. W., P. S. Nieberg, and R. J. Buhr. 1990. *Poultry Sci.*, 69:1553.

Wismer-Pedersen, J. 1957. *Proceedings of the 4th European Meat Researchers Workers Conference*, 18:3.

Wismer-Pedersen, J. 1959. *Food Res.*, 24:711.

Xuewei, L., X. Yongzuo, Z. Junshen, L. Fangqiong, L. Shijian, and Z. Li. 1992. A comparative study on meat and fat quality of Sichuan, exotic and crossbred pigs, in *Proceedings of the International Symposium on Chinese Pig Breeds*, C. Runsheng, ed., Harbin, China: Northeast Forestry University Press.

Zhang, W., D. L. Kuhlers, and W. E. Rempel. 1992. *J. Anim. Sci.*, 70:1307.

CHAPTER 3

Optical Probes

INTRODUCTION

ON-LINE PROBES FOR the evaluation of meat quality may come in all shapes and sizes, as seen in Plates 1 and 2, and no technology or operating principle need be excluded, provided that it produces commercially or scientifically useful information. Optical probes are well established, commercially important, and technically very interesting — so they warrant their own chapter.

The idea of evaluating meat on-line objectively first became established with the discovery that probing the electrical properties of meat could produce commercially useful information (Chapter 6). Impedance or conductivity measurements were used initially to measure PSE, but then the technology was applied to another on-line application, measuring subcutaneous fat depth. Later, it was found that optical probes are better at measuring fat depth than electrical probes, and this led to the well-established market for optical fat-depth probes that we see today, what we might call phase 1 of the on-line evaluation of meat. Phase 2, which is starting right now, stems from the realization that optical probes can do much more than simply measure fat depth.

The original idea for an optical probe to look within the meat of the carcass seems to have come from work on live pigs in the 1950s: a scalpel was used to make an incision through the back of the live pig to measure its fat depth (Hazel and Kline, 1952; Aunan and Winters, 1952). This crude principle was adapted for early optical probes by using a light pipe, a glass rod covered with a silvered, mirror surface facing inwards. Light that enters one end of the pipe is reflected to the other end. Thus, the fat to muscle boundary appearing at a lower window on the light pipe within the animal or carcass may be seen by the observer looking through the top window, and then the depth of the bottom window in the carcass may be determined from ruler markings on the outside of the probe.

Automation of these operations led eventually to the fat-depth probes that now are used internationally for pork carcass grading (Chapter 1). Essentially, the bottom window of the light pipe was replaced by a light-emitting diode (LED) and a detector. When the optical window subtended by the diodes passes from the muscle into the fat, as the probe is withdrawn from the carcass, the detector responds to an increase in reflected light. Simultaneously, the depth of the probe is read from a device connected to a plate on the surface of the carcass, so that a microprocessor may use these inputs to find the depth of back-fat and then estimate the meat yield. The electromechanical aspects of probe design warrant separate consideration (Chapter 4).

It is important to acknowledge the work of Kapany (1967), who pioneered the medical and biological applications of fiber optics. Another major step in the development of optical probes for meat quality was undertaken by MacDougall (1980), who replaced the light pipe of the early fat-depth probe with two fiber-optic light guides, one taking light into the meat and the other gathering scattered light to be passed back through a filter to a photometer (Figure 3.1). For the future, we need to look at all available wavelengths and the use of multiple fiber light guides.

Figure 3.1 *A sketch of MacDougall's TBL mark II fiber-optic meat probe for the measurement of PSE and DFD with minimum interference from myoglobin (TBL Fibre Optics Group, Leeds, UK).*

In summary, looking at fat depth with a probe is operationally related to looking at meat quality with a probe. Starting with a few commercially advantageous operations, such as probing for back-fat and for PSE or DFD, we may begin to see a new technology emerging—the optical probe. The objective in this chapter is to introduce its methodology. Figure 3.2 provides a key to symbols used to keep block diagrams as simple as possible: PD is used to denote a photodiode, LED for a light-emitting diode, and a question mark for the sample being measured.

PROBE COMPONENTS

Light Source

For a simple application such as a fat-depth probe, the meat may be illuminated from an LED mounted in the shaft of the probe, but white light is needed for colorimetry, and intense light is needed for other applications. Thus, a relatively large light source may be needed, either mounted in the body of a handheld instrument or at a remote location near to the microcomputer controlling the system. Figure 3.3 shows the emission spectra of some common light sources. The smooth spectrum for the 12 V, 100 W filament lamp (line 1 in Figure 3.3) is the most useful for colorimetry because there are no sharp emission peaks, as in the other two spectra. Intense peaks cause major problems when scanning measurements are made across the spectrum. If the photometer is set to a maximum reading on the most intense peak, then the light intensities in the valleys are so low that only a small fraction of the dynamic range of the photometer is used—rather like using a thermometer with a range from 0 to 100°C to look for small differences in temperature between 1 and 2°C. Also, small errors by a monochromator in arriving at a wavelength will cause a major variation in light intensity if the wavelength is near an intense peak.

The filament source (line 1 in Figure 3.3) has very little intensity at 400 nm, which may be only a minor problem in reflectance colorimetry of meat reflectance, because the contribution of violet light to human color perception is relatively small. Violet light is important for measuring the blue whiteness of fat but not for the redness of meat. However, light scattering causes some interesting effects when light passes through meat (Chapter 5), and a strong source of low wavelengths is necessary

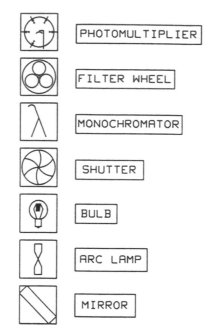

Figure 3.2 A key to symbols for optical components.

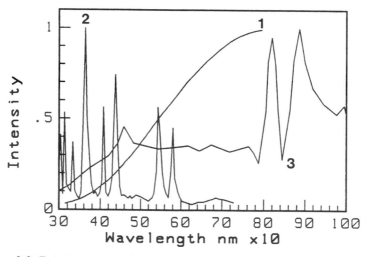

Figure 3.3 Emission spectra of three light sources: a 12 V, 100 W filament lamp (1), a short-arc mercury lamp (2), and a short-arc xenon lamp (3).

to measure these effects, some of which may have a potential commercial importance.

Line 2 in Figure 3.3 is the emission spectrum for a short-arc mercury lamp, which is familiar to anyone who has used a UV fluorescence microscope. It has some very sharp, intense peaks that make it virtually impossible to use as a source for conventional spectrophotometry, but many of these peaks are ideal for exciting fluorescence. One of the peaks, at 365 nm, is very close to the excitation maximum of collagen in meat, thus providing strong excitation of connective tissue fluorescence in meat.

Line 3 in Figure 3.3, for a short-arc xenon lamp, is somewhat of a compromise between the first two emission spectra. It has useful high intensities around 400 nm, a relatively flat plateau across most of the visible wavelengths, and some major emission peaks in the near infrared that may be exploited or avoided, depending on the application.

There are other important factors apart from the emission spectrum to be taken into account. Short-arc lamps are ideal for launching high light intensities into optical fibers, but they get very hot. Heat-absorbing or reflecting filters may be required to protect optical fibers, certainly if they are plastic fibers, and it is difficult to mount an arc lamp in a hand held unit for on-line use. Thus, arc lamps can only be used easily in laboratory research or for on-line applications if the lamp has been placed in a secure, remote location, usually near the computer. Given that several thousand volts at a high frequency are required to start the arc, before it switches over to a relatively low-voltage DC maintenance current, it is very difficult to use an arc lamp safely in the humid conditions of the meat plant. Also, arc lamps are short-lived, expensive (amortized at several dollars per hour to operate), and fragile; have limitations on the angles at which they can be tilted during operation; and require a stabilized power supply.

Lasers may be extremely useful sources of coherent monochromatic light in laboratory research on meat, as in measuring sarcomere length by diffraction, and they may have some on-line applications, as proposed by Birth et al. (1978) and considered in Chapter 5.

Monochromator

In Figure 3.2, λ stands for wavelength, to indicate a filter or monochromator. Those used in on-line apparatus are likely to be fairly simple filters with a wide bandpass. The simplicity keeps the cost down, the wide bandpass enables a relatively high light intensity, and the choice of wavelength already has been decided during laboratory development of the system. In the research laboratory, however, the place of the single filter in the final system is likely to be taken by a grating monochromator or a prism monochromator (the latter being the older technology, more readily available as surplus). This allows for the flexibility required to optimize a system. For example, reflectance measurements might be made across the spectrum and correlated with a feature of commercial importance, such as water-holding capacity or emulsifying capacity. The peak wavelength with the highest coefficient of correlation then would be chosen for a fixed filter in the final on-line system.

The efficiency of a monochromator at different wavelengths is important. Figure 3.4 shows the efficiencies of two different gratings, one suitable for UV fluorescence of connective tissue in meat and the other for colorimetry. The differences result from the design of the grating in terms of the groove angle, shape, and depth, with the wavelength of peak efficiency around which the design is based being known as the blaze wavelength.

One of the components symbolized in Figure 3.2 is a filter wheel. This may be used to rotate filters through the light path, allowing a single photometer to obtain information at a few different wavelengths, but a filter wheel also is required in conjunction with a grating monochromator. The diffraction grating monochromator passes a series of spectra

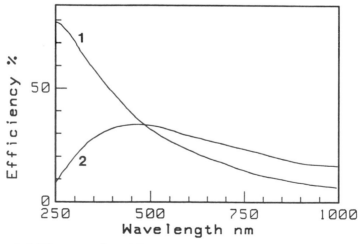

Figure 3.4 *Efficiencies of two diffraction grating monochromators, one for UV (1) and the other for visible light (2).*

with different orders of diffraction, and these create harmonics. For example, although the nominal output of a grating monochromator (the first-order diffraction spectrum) might be 200 nm, the first harmonic might be at 400 nm. So, instead of UV fluorescence in the meat, we could be measuring violet reflectance. To avoid the error in this example, the filter wheel would be set to remove any light at 400 nm. A filter wheel is required because several filters, usually three, may be required to safeguard against this error across the whole spectral range of the monochromator. This problem is very important in transferring a technique from the laboratory to an on-line application. Solenoid-operated swing-in filters may serve the same function as the filter wheel, and they are much faster and allow multiple filter combinations.

A conventional grating or prism monochromator has input and output slits in the light path, and there should be a sharp image of the input slit at the output slit. With the advent of diode array detectors mounted on a single chip, another monochromatizing device has become common, the spectrograph. The optics are altered so that the array detector may be located directly on the instrument where the output slit once was, and images of the input slit are dispersed across the array. Thus, unwanted wavelengths (stray light) from reentrant spectra reflected from within the optical system must be eliminated at the source, as in the Czerny-Turner design spectrograph.

In transferring technology from the laboratory to an on-line application, it may be important to allow for a decrease in performance. For example, data from a set of spectral measurements made on a research-grade monochromator or spectrograph might be used to develop a prediction equation, but this might fail when used for on-line data collected at a low resolution. A broadband simple correlation, although less impressive in the laboratory, might be more robust and survive to give better results on-line.

Shutter

A shutter is a useful tool for standardization and fault-finding in a laboratory system but may be omitted from an on-line system for cost reasons. The shutter symbol in Figure 3.2 is for an iris shutter, as used in a camera. Iris shutters are very fast, and the segments withdraw symmetrically from the light path, which may be an important advantage over a less expensive solenoid-operated swing-in shutter that moves across the light path from one side to block it. The three primary applications of a shutter are to find a photometer reading in total darkness, to find the response time of a photometer, and to protect a specimen from continuous illumination.

The dark-field photometer reading is fairly straightforward: it has to be subtracted from the photometer reading for the sample being measured, just as a weigh-scale has to be tared for the weight of the container holding the meat. If the light on the photometer is quite intense, the dark-field reading may be zero, or very close to it. But, typically, what happens is that the gain of the photometer must be increased to measure low intensities of light, and then there is a residual measurement detectable from a dark field. For on-line applications, the place of the shutter may be taken by an instruction to the user to place a light trap over the probe and to obtain a measurement when no light reaches the photometer, often called the dark calibration. In a research application, it is convenient to do this automatically with a shutter to check on the performance of the system.

Finding the response time of the photometer is only required in the research laboratory: by the time a system is used on-line, all the appropriate decisions about photometer response time should have been made. The main application is in developing probes that work on the fat-depth probe principle, taking measurements as the probe moves through the carcass. If the photometer response time is slow, relative to probe velocity, it will smear or miss the incoming data as the probe window passes the fat to muscle boundary. Figure 3.5 shows some examples of a photomultiplier responding to flashes of light created by an electrically triggered camera shutter.

Photometer

There is no such thing as a simple photometer. The ones that look simple, like the photodiodes for sale in a hobby shop, have complex electrical characteristics (junction capacitance, shunt resistance, and series resistance). Paradoxically, complexity in the control circuitry is required to make the output of a photometer a simple, linear function of light intensity. This may be a problem in transferring technology from the laboratory to an on-line application, which is likely to involve a switch from simplicity achieved by complexity to complexity resulting from simplicity (and low cost). Although we may leave the electrical problems to a professional engineer, the key point the meat researcher must grasp, as with the light source and the monochromator, is the effect

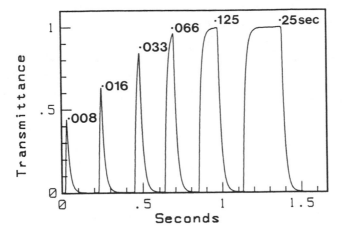

Figure 3.5 *Photomultiplier responses to flashes of light created with the iris shutter of a camera (Swatland, 1993a).*

of wavelength on a photometer. Figure 3.6 shows the spectral response of a typical silicon photodiode, poorly suited for UV measurements but with a fairly flat response across the visible and near-infrared.

Photodiodes are ideal for on-line use because of features such as stability, low cost, and rugged construction, but there are other possibilities, including photomultipliers and photodiode arrays. Photomul-

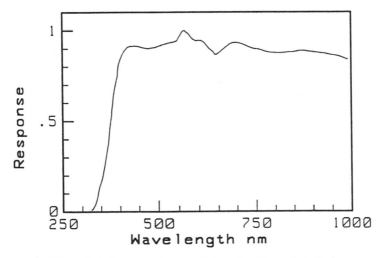

Figure 3.6 *Response characteristics of a silicon photodiode.*

tipliers are well established technology, with books written about how to use them and a ready supply of parts in surplus scientific apparatus. Contained in a vacuum tube are a photocathode, an electrostatic focusing system, dynodes that act as multipliers, and an anode. After light strikes the cathode, electrons are released, the current is multiplied by the dynodes (at successively higher positive voltages), and the current from the anode is fed to an amplifier. The metallic composition of the cathode has a major effect on spectral sensitivity, allowing several major types for different applications. These are designated as S20 (cathode composed of Na, K, Cs, and Sb), S1 (Ag, O, and Cs), S5 (Cs and Sb), and gallium arsenide. The window for the light may be on the side of the tube (allowing space for nine dynodes) or at the end of the tube (to accommodate fourteen dynodes). Miniature photomultipliers also are available.

Diode arrays, which used to be exclusive to astronomers and physicists with massive government funding, have fairly recently become available to the rest of us with more humble aspirations and funding. A spectrograph with a diode array can do the same thing as a monochromator with a stepper motor and a photomultiplier, but almost instantaneously and without moving parts. Self-scanning photodiodes are supported on a single integrated circuit, each diode with its own storage capacitor to integrate the current, which then is scanned to the output line. Thus, unlike the photomultiplier, which measures light intensity in watts, the diode array measures energy over time, joules. The Colormet probes seen in Plates 5 and 6 are based on diode arrays, which have many features ideal for on-line use. Relative to other components such as vacuum tube photomultipliers, diode arrays are small and compact, do not require high voltage circuitry, and are rugged and reliable. Of particular importance is their high speed of operation, which enables keeping pace with the fastest line speeds in the meat industry. Spectra can be collected at 10 to 20 Hz, thus allowing real-time spectrophotometry of meat slurry.

Optical Fibers

On-line evaluation of meat is very much an interdisciplinary subject, where professionals from diverse fields such as computer science and electrical and optical engineering may react with their counterparts in the technical and scientific side of the meat industry. The meat scientist needs at least a general idea of what the engineer is doing and vice versa. But the interface between optical systems and meat, quite literally, between optical fibers and muscle fibers, is a special case. Here, neither side really understands what is going on, and nobody is in charge. Engineers know a lot about the performance of optical fibers, and biophysicists know a lot about the optical properties of muscle, particularly in relation to the molecular biology of muscle contraction, but very little is known about the interface between optical fibers pushed into the muscle fibers of meat. Thus, before we look in detail at the optical properties of meat (Chapter 5), some knowledge of optical fibers is desirable.

Figure 3.7 shows a transverse section (TS), a refractive index (RI) profile, and a longitudinal section (LS) of a step-index multimode fiber. The core of the fiber (diameter d_1 and RI $= n_1$) is surrounded by cladding [thickness $(d_2-d_1)/2$ and RI $= n_2$] that causes total internal reflection to conduct light along the core. A bare core still conducts light if the surrounding medium has a lower refractive index and the primary purpose of the cladding is to improve efficiency and prevent short circuits. The refractive index profile shows the step index change, from n_1 to n_2, where the cladding meets the core.

In the longitudinal section of the optical fiber in Figure 3.7, a ray of monochromatic light enters the core at angle α from a medium beyond the fiber, having a refractive index $<n_1$. The ray is refracted towards the normal (broken line) as it enters the optically denser medium of the core. If angle α is increased beyond the maximum coupling angle (α_{max}), rays no longer enter the core and are reflected from its surface. The numerical aperture (NA) of the fiber is given by

$$NA = \sin \alpha_{max} = \sqrt{(n_1^2 - n_2^2)}$$

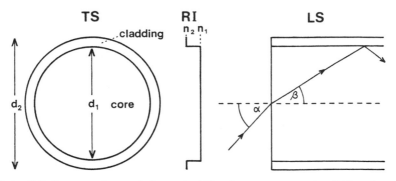

Figure 3.7 *Features of a step-index optical fiber in transverse section (TS), refractive index (RI) profile, and longitudinal section (LS).*

Rays with angle $\cos \beta < n_2/n_1$ are guided along the core, but rays with different angles of entry have different path lengths or modes along the core. A high numerical aperture and a large angle β both allow more modes than a low numerical aperture and a small angle β.

Multimode dispersion generally is not a problem in on-line evaluation of meat (unless using pulses of light or interferometry), so step-index fibers are satisfactory. Suitable fibers may range from $d_1:d_2 = 100:140$ μm, used in bundles of about fifty fibers for spectrophotometry, to $d_1:d_2 = 1000:1050$ μm, used for single-fiber fluorometry.

Monomode fibers are step-index fibers with a very narrow core. This makes coupling between optical fibers difficult but only allows a single mode to propagate, so that mode dispersion is low. In graded index fibers, the refractive index of the core increases parabolically towards the central axis of the optical fiber so that modes travel in sinusoidal waves. Modes with long light paths through the outer parts of the core travel faster than those with short paths near the center of the core, so that differences in path length and speed tend to cancel each other, and mode dispersal is low. The acceptance angle of graded index fibers is high at the axis of the fiber and low near the cladding boundary.

The coupling of optical fibers always requires care and attention because small alignment errors may produce large changes in coupling efficiency. Coupling efficiency seldom is very high and depends on wavelength. One of our major problems in using optical fibers for on-line evaluation of meat is that the coupling efficiency of optical fibers to muscle fibers is subject to numerous sources of variation. The other major problem is that there is a fundamental difference between coupling optical fibers to meat and coupling them to a typical optical standard, such as a white plate. Between optical fibers and muscle fibers, there is a direct junction between plastic or quartz and the meat fluid, and this is difficult to simulate when coupling with the optical standard. A judicious standardization protocol may be used to correct for the effect of wavelength on coupling efficiency, but this must be tested by proving that a blank sample will return a flat spectrum at all wavelengths. This is more difficult than it first appears, because the reflectance standard must be coupled to the optical fibers using a medium of the same refractive index as that in the fluid phase of the meat. Meat is a particularly difficult commodity in this regard because a decrease in pH decreases the negative electrostatic repulsion between myofilaments. This releases fluid with few solutes and, hence, a relatively low refractive index, which, in turn, may change the coupling efficiency at the meat to optical fiber interface.

Light Guides

Anything that guides light along a complex circuit may be called a light guide, such as one or more optical fibers, a glass rod with a silvered surface, or even inside a jet of water — as in John Tyndall's first scientific demonstration of the effect in 1870 (although Venetian glassblowers had long been exploiting internal reflectance in ornate glassware). Light guides may be fashioned into a variety of configurations for on-line use, the simplest and most important being a splice where two or more bundles of optical fibers are bound together in the same protective sheath to form a bifurcated light guide.

At the window of the common trunk, ingoing optical fibers to the spectrophotometer may be separated from outgoing illumination fibers in three basic ways:

(*1*) Each on one side of a diameter across the window to give two D-shaped areas

(2) Concentrically, with one group of fibers in a ring around a core of the other type

(*3*) Randomly arranged in the window

The first of these is the easiest to construct in the workshop, while the second is not too difficult if a layer of foil is used to bind the central core fibers. The third, a random pattern, looks simple at first, but achieving true randomness is very difficult (because each bundle of fibers coming in from one side tends to bunch together). Scattered is probably a more accurate term than random. The common trunk may be kept short to limit the cross-talk between ingoing and outgoing fibers (for example, when the outgoing illumination is very intense and the ingoing returned light is very weak). A foil partition between the core and ring fibers of a concentric arrangement also may be used to limit cross-talk.

A concentric arrangement allows a bifurcated light guide to be used with a mirror to measure absorbance in a liquid. The mirror is placed at a fixed distance, parallel to the window of the light guide. Light from outgoing fibers of the light guide core is reflected back into the outer ring of ingoing fibers, which enables the optical path length to be fixed.

The ratio of ingoing to outgoing fibers may be varied to optimize the performance of the system: in most cases, more ingoing fibers than outgoing fibers are required since coupling with a monochromator via a rectangular slit is less efficient than coupling with an illuminator focused to a point. A ribbon converter, where the optical fibers of one branch are lined up to form a rectangular window, improves the coupling to monochromators and spectrographs, which generally have elongated rectangular entrance slits.

Individual optical fibers may be coupled in many different ways for small-scale laboratory work. Fibers may be fused carefully with a heat gun or soldering iron after they have been twisted together, so that the cores of the fibers communicate. Unused endings are sealed with a light trap. Optical fibers may carry stray high-order modes in the cladding, especially after passing through a fused coupling to another fiber. High-order modes may be removed by passing the naked core of a fiber through a small light trap filled with high refractive-index oil.

APPLICATIONS

Fat-depth Probe

The optical components of a Danish fat-depth probe are shown in Figure 3.8. Light from an LED at 950 nm leaves the optical window of the probe and is reflected by carcass tissues back to a photodiode (PD). A reference (R) comparison is obtained from another photodiode via a mirror reflection within the shaft of the probe. This follows the well-proven laboratory principle of including a reference alongside a measured sample, as in a dual-beam spectrophotometer. Thus, the analysis of the result is based upon the relatively stable ratio of the sample to the reference, rather than on absolute measurements of a single sample source easily influenced by the operating conditions. Within reason, therefore, the PD output intensity has little effect on the measurement, which is the ratio of meat reflectance to the internal mirror reflectance.

Spectrophotometry of Meat Products

NIR spectrophotometry is a widely used and certified technique for food analysis (Norris, 1984). Special equipment is commercially available for the analysis of plates and dishes of comminuted meat products.

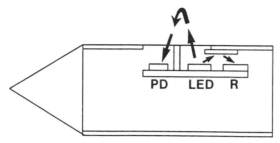

Figure 3.8 *Optical components of a Danish fat-depth probe (Thodberg, 1993), showing a light-emitting diode (LED), a photodiode (PD) detecting meat reflectance, and a reference (R) photodiode monitoring the output of the LED from a mirror reflection within the shaft of the probe.*

Applications include the prediction of the functional properties of batches of raw meat ingredients used for further processing, checking the nutrient composition of products, and maintaining fat levels at specified tolerances in ground meat, especially beef hamburger. NIR spectrophotometry also may be used to predict taste panel scores for beef tenderness (Hildrum et al., 1994).

Typical optical components described by Workman and Andren (1992) for NIR include a tungsten-halogen light source, a grating monochromator (800−1100 nm with a 6-nm bandpass), and a mirror and lens to direct the light 15−20 mm through a plate of comminuted meat (Figure 3.9). Multiple scanning and rotation of the sample are required to average its heterogeneity, and smoothed absorbance measurements at 100 wavelengths are used to make predictions from a partial least squares chemometric calibration. Problems are caused by drying of the sample surface, pH-related changes in light scattering (Chapter 5), and metmyoglobin formation (Chapter 7), but sample temperature may be corrected.

Conway et al. (1984) extended the optical components for NIR spectrophotometry using a bifurcated light guide so that measurements could be made on human patients to monitor adiposity, as shown in Figure 3.10. This arrangement is vulnerable to ambient illumination reaching the photodiode, so the probe on the patient's skin was protected with black felt. Scanning from 700 to 1100 nm, strong correlations were found with other methods of measuring human adiposity: deuterium oxide dilution ($r = 0.94$), skinfold thickness ($r = 0.90$), and ultrasound ($r = 0.89$). Unfortunately, thick hides with bristly hairs prevent this method from being used on live meat animals, but it is applicable to meat

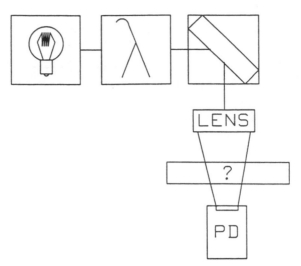

Figure 3.9 *Optical components of an NIR spectrophotometry system for on-line measurements of fat, moisture, protein, and collagen in comminuted meat products (Workman and Andren, 1992).*

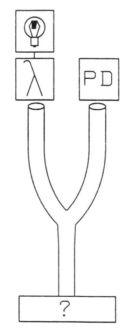

Figure 3.10 *Optical components for fiber-optic spectrophotometry of a human subject (Conway et al., 1984) or a sample of meat (Swatland, 1982).*

products (Swatland, 1982) and gives good results for predicting fat content (Eichinger and Beck, 1992). The basic problem is not in the optical components, but in averaging sample heterogeneity. Locating a monochromator in the outgoing pathway, rather than along the ingoing illumination pathway, is less vulnerable to ambient illumination.

Spectrophotometry of visible light is particularly useful for the measurement of meat color (Chapter 7). Measurement of meat pigments on-line is commercially advantageous (Andersen et al., 1993). The Japanese TS-200 is suitable for measuring meat pigments in meat and in model systems for food products (Tsuruga et al., 1994).

Detection of PSE and DFD

MacDougall's (1980) fiber-optic probe for the detection of PSE and DFD (Figure 3.1) has simple optical components, which is somewhat of a virtue where cost and reliability are concerned. But for a wider range of applications, such as measuring poultry, fish, pork, veal, and dark-cutting beef on-line, as well as fat color and functional properties of processed meat products, some elaboration is required. The Colormet meat probe (Plates 5 and 6) has a photodiode array (PDA in Figure 3.11) on a prism spectrograph, and the PDA is synchronized with a xenon flash unit (Xe in Figure 3.11). A reference to correct for variation in flash intensity is obtained via a small light guide from the flash to a photodiode. The original prototype had a probe with a flat ending, but this was changed to a ring of fibers opening around the shaft of the probe in the final model. The scattered arrangement of illumination and receiving fibers was chosen to optimize the detection of light scattering in PSE pork, which raises a difficult question.

Our current theoretical understanding of the optical properties at the junction between optical fibers and tissues is rudimentary, to put it politely. Thus, we may build probes that empirically show a good response, either to myoglobin in solution between the scattering microstructure of meat or to the scattering microstructure. Thus, probes that excel at myoglobin detection in veal grading may have a poor performance for detecting PSE or DFD and vice versa.

Connective Tissue Fluorescence Probe

Connective tissues in meat are fluorescent, emitting blue-white light when excited by UV light at 365 nm (Chapter 5). The principle of the

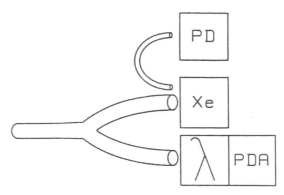

Figure 3.11 *Optical components of the Colormet meat probe developed by Ken Butt (Swatland, 1986).*

fat-depth probe, to make a series of measurements through a small probe window relative to depth in the meat (Figure 1.4), was adapted to monitor the distribution of connective tissues in meat. Thus, instead of searching for a discrete boundary between fat and muscle, the connective tissue probe passes through the muscle detecting connective tissue fluorescence in relation to meat depth (Chapter 9).

It is not possible to mount the light source and detector within the shaft of the UV probe, as seen in Figure 3.8. Instead, a single optical fiber is used to create a window on the shaft of the probe; then, back near the main controller, the source and detector light paths are separated by a dichroic mirror (Figure 3.12). At wavelengths less than 400 nm, the light from the mercury arc is reflected from the dichroic mirror into the optical fiber, passing along the fiber and out of the probe window.

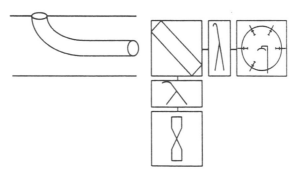

Figure 3.12 *Optical components of a UV probe for connective tissues in meat (Swatland, 1991a, 1991b). The curved optical fiber opens at a window on the side of the probe shaft.*

Fluorescent connective tissues at the probe window emit light that is collected by the optical fiber and passed back towards the controller. Instead of being reflected into the illuminator, most of the fluorescence, at wavelengths over 400 nm, passes through the dichroic mirror to the photomultiplier. Filters are required in front of the arc lamp to remove some of the heat and most of the light over 400 nm, while a high-pass filter in front of the photomultiplier helps prevent any stray light below 400 nm from reaching the photomultiplier. This relates to the phenomenon of pseudofluorescence from meat (Chapter 8).

The components for a carcass probe (shown in Figure 3.12) may be adapted for on-line use in meat processing by replacing the optical fiber with a 1-cm diameter quartz rod. Thus, instead of pushing the optical window through the tissue of the carcass, the probe window opens into a slurry of meat, which might be moving. This enables the skin or connective tissue content of meat processing slurry to be determined in batch or continuous mode (Chapter 12).

Laboratory System

One of the special advantages of optical fibers is that they enable the analytical power of a research laboratory to be connected directly to an on-line meat industry situation. Thus, using research-grade apparatus in a remote laboratory, it is possible to make many types of optical measurements on a kill floor or in a meat cooler or processing plant. Ellipsometry and interferometry are difficult at present, but colorimetry, absorbance spectrophotometry, and fluorescence measurements are relatively straightforward. A general purpose laboratory system is shown in Figure 3.13.

A rotating mirror allows light sources to be changed readily. For example, a filament lamp with an emission spectrum similar to that shown by line 1 in Figure 3.3 might be used for colorimetry or spectrophotometry, then switching to a mercury arc (line 2 in Figure 3.3) to measure the fluorescence emission spectrum of connective tissue in meat. A shutter is required on the primary illumination pathway, both to find the dark-field current of the photomultiplier and to shut off the primary illuminator if necessary. For example, a known source of white light supplied independently is necessary for calibration in many fluorometry operations. Absorbance spectrophotometry and reflectance colorimetry usually are based on the measurement of ratios (light into a system versus light out), but fluorometry is not normally a ratio and

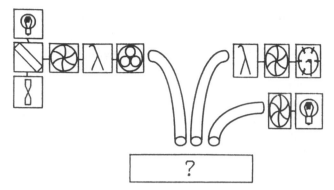

Figure 3.13 *A versatile laboratory system connected via optical fibers for on-line measurements.*

requires the photometer to be shown a white source with a known emission spectrum, against which it may evaluate the fluorescence being measured. In Figure 3.13, the secondary illuminator for this purpose is connected to the sample by its own light guide.

SOFTWARE

When developing software for on-line applications, it is advantageous to anticipate as many future applications as possible. With foresight, software may be developed to collect and analyze data from many different types of systems, thus minimizing delays, providing the maximum return on any software investment, and helping to clarify the logical design of on-line measurements. With a top-down approach to programming, the underlying similarities of a wide variety of on-line measurements become apparent.

Graphics Requirement

Intuitively, the first way to examine a spectrum or an optical signal is to look at it, and this requires graphics software. A variety of methods may be used afterwards to obtain further information, but the graphics software is at the junction between the initial acquisition and the later processing of data. It is important, therefore, that this junction should never become an obstacle to innovation, and a primary consideration is to anticipate the variety of forms in which incoming data may appear.

A standard data format is preferable for all types of data to be graphed by a research group or quality control department, but this format must have maximum flexibility and also be convenient for data storage. Eventually, thousands of spectra may be stored together, and they will be of little use unless they carry coded information on how they were obtained and what they describe. The same graphics display as used for photometric data should also be adaptable to statistical analysis, such as plotting a correlation coefficient in place of a reflectance measurement on the *y*-axis.

Data Format

Unfortunately, the optical data collected on-line may be irregular, for example, differing in the number of wavelengths measured for each spectrum. Furthermore, wavelengths are not the only domains that may be scanned. Birefringence could be an important optical attribute of meat for future measurement (Chapter 5) and is measured by angle rather than wavelength. However, the operations involved in angular scanning are very similar to those involved in spectral scanning and may be controlled with the same software using a common data format. In other words, there is little difference operationally between incrementing a monochromator and taking a measurement and incrementing a polarized light analyzer and taking a measurement.

Problems may occur if angles, rather than wavelengths, appear on the *x*-axis of a graphics display. For example, it might be necessary to scan from $-45°$ to $+45°$ each side of a setting at $10°$, in which case the scan will be from $325°$, through $360°$, to $55°$. In other words, it should not be assumed automatically that the correct order of *x*-axis data always forms an ascending series.

Spatial scanning is another possibility for the *x*-axis. This involves incrementing the position at which a measurement is made to scan across or through a meat structure. An example might be pushing an optical probe through a batch of ground beef to integrate measurements of the fat content or metmyoglobin formation. An active scanning device, where each increment in position follows a programmed sequence, produces a series of measurements at regular intervals, whereas a handheld carcass probe such as a fat-depth probe may produce irregular data. If the motor force originates from the human arm, a passive scanning device such as a fat-depth probe may be triggered either at set increments in position or in time. The latter option produces a series of measurements that are not spaced at regular intervals in position. Thus, it is difficult to process irregular vectors such as these if the x-axis has been reduced to its determinants (minimum, increment, and number of values), which might otherwise be done to save on storage space for a regular spectrum.

Top-Down Programming

Bearing in mind the operational similarities in many different types of optical scanning (wavelength, angle, and position), it is possible to program a hierarchy of primary, secondary, or multiple scanners. For example, to make a series of ellipsometry measurements along a transect through a sample, the primary scanner might be a rotary analyzer with a stepper motor as a secondary scanner. This is radically different to using excitation and emission monochromators for fluorometry of connective tissue in meat, but the method of data collection and the resulting data matrix are both very similar. Determining the order of precedence for scanners may depend largely on the apparatus involved and requires an understanding of the speed and accuracy of each scanner.

In subprograms for operating scanning devices, the primary algorithm generally includes a "do and wait" operation (for a stepper motor) or a "do and look" operation (for a servo system). Software must be written to be independent of controller speed. For example, with a new controller running at a faster clock speed than an old one, the faster "do" operation will require a longer "wait" if the scanner operation is simply matched to the duration of "do + wait."

If properly programmed, a general purpose system, such as that shown in Figure 3.13, may take a variety of different types of measurements in rapid sequence once the optical fibers have been located in the meat sample. Flexibility is more easily attained centrally in the research laboratory than out on the kill floor or in the processing plant. It takes time and effort to get a probe into a unit of meat, so it makes sense to optimize the acquisition of information once this has been done.

OPTICAL WINDOW

In many aspects of the on-line evaluation of meat, we may adapt or adopt a wide variety of optical components, backed up by commercial suppliers, consultants, and academic knowledge, but when it comes to the optical window, the interface between the meat and the optical system, we are on our own. There are many aspects to the meat-optics interface, and much of the basic thinking and experimenting has not yet been undertaken.

Vignette Window Effect

If the image of a beef carcass is projected onto a vidicon tube or CCD camera (Chapter 11), the whole pattern of muscle shape and sub-cutaneous fat distribution may be captured in a single frame. But if a fat-depth probe is pushed into the same carcass in search of similar information (muscularity and fat distribution), the information is acquired sequentially through a small window. The small window may only give a vignette of anatomical structures that it passes. Imagine a person riding in a railroad boxcar and peering out through a crack in the door: the scenery is chopped into a series of light and dark impulses, and a thick telegraph pole starting to pass close to the door might initially look the same as the start of a tunnel. In other words, a small optical window vignettes or obscures passing structures that are larger than the window.

This window effect may be compounded if there are differences in the luminous intensity of objects that pass the window. Consider another analogy: the sun is shining through the window of a room onto a large photometer, and the window has a roller blind of opaque material that may be drawn down to block the light coming through the window. By experiment, it is possible to find a photometer reading that corresponds to the position of the blind halfway down the window. This reading then may be used to identify the halfway position, so that a computer may be programmed to recognize the "blind-up" and the "blind-down" positions. Essentially, this is the way in which an optical fat depth probe finds the boundary between the back-fat and the longissimus dorsi (Figure 1.4).

PSE is the major problem for optical fat-depth probes used in pork grading. Returning to the analogy of the window blind, if a translucent blind is substituted for the opaque blind, the photometer gives biased readings for finding the position of the blind, because the translucent blind is more than halfway down before the photometer threshold (half light intensity) is reached. Similarly, the algorithm in the fat-depth probe may fail if the muscle is PSE and has a similar reflectance to the fat (if $F_{max} = M_{min}$ in Figure 1.4). Computer neural networks are used in the Danish Carcass Classification Centre to minimize this problem (Thodberg, 1993).

The window, or optical aperture, of the probe might be made as small as possible relative to the expected fat depth, but this may cause other problems. First, the small window will be sensitive to small structures such as blood splashes in the fat and marbling fat in the lean (Figure 10.3). Secondly, the amount of light reaching the photometer will be reduced, thus requiring greater amplification, which, in turn, will increase the noise of the signal. As the noise level rises, it will become progressively more difficult to locate the fat to muscle boundary. To overcome the problem of a low light intensity at the photometer, the intensity of the light source might be increased. But then the photometer reacts to an enlarged cone of diffuse light outside the window, so that the operational window size is increased again, failing to solve the problem.

The problem might be minimized if the illumination window and the photometer could be placed side by side, perpendicular to the long axis of the probe, so that both the light source and the photometer window cross the muscle to fat boundary together. But there is not enough space to mount the diodes in this way without increasing the diameter of the probe to the point that it causes artifacts when pushed through the meat.

The optical windows in meat probes are complex structures, as shown in Figure 3.8, where the magnitude of the vignette window effect is a function of the sum of the two windows along the length of the probe. To understand the basis of how such a window will respond, we may start with some simple geometry.

In Figure 3.14, the optical window formed by an optical fiber is shown by the ellipse on the left. There tends to be a trade-off with optical fibers: the soft, plastic fibers that can sustain a small radius of curvature are limited in the wavelengths they transmit, while the relatively rigid quartz

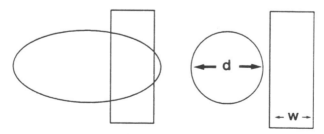

Figure 3.14 *Geometrical models for signals created at an optical window. Windows on meat probes often form an ellipse (left), but these can be modeled as a circle (diameter, d), which is passed by a rectilinear bar (width, w) that represents anatomical structures such as seams and fibers of connective tissue in the meat.*

fibers with high transmittance at useful wavelengths for meat research can only be used at a large radius of curvature. Thus, a side-mounted window in a probe is likely to be elliptical, although it may be modelled as a circle, as shown on the right of Figure 3.14. Similarly, the geometry involved applies equally well to a group of light guide fibers that collectively form a window with an elliptical or circular outline. For further simplification, it is assumed that a single optical window acts both to illuminate and measure, as shown in Figure 3.12.

If the window is circular with a known diameter (d) and the complexity of anatomical shapes passing the window is modeled by a rectilinear bar with a known width (w), then a photometer that measures the reflectance (r) of the bar through the window will produce the signal shown by line a in Figure 3.15 for $d = 1$, $w = 1$, and $r = 1$. The signal rises as the leading edge of the bar enters the window, reaches a maximum when the bar fills the window, and declines as the trailing edge of the bar leaves the window.

Comparison of lines b and c in Figure 3.15 shows that, by itself, the amplitude of reflectance cannot be used to distinguish between bars that differ in both width and reflectance. Furthermore, a bar with $w = .5$ and $r = 1$ attains a reflectance $> .5$ because it fills more than half the window area when the center of the bar is at the center of the window (line d in Figure 3.15). Thus, with real signals from complex anatomical structures in the carcass, both the amplitude of the signal and a measure of its peak width are required to resolve the anatomical feature passing the window.

A variety of factors affect the shapes of real signals. Apart from the complex shapes of many anatomical structures such as seams and fibers of connective tissue, smearing may occur across the optical window as the tissue is deformed or disrupted by the probe. The response time of the photometer also may produce a similar distortion of signal shape as a function of probe velocity versus photometer response time. This is why data such as those shown in Figure 3.5 are important. Bearing in mind that the probe moves in a straight line through the meat, the most important dimensions of the window and of the measured objects are their effective diameters parallel to the probe direction.

Predictions may be tested experimentally by measuring the reflectance of white bars against a black background, as a test chart moves past the optical window of a meat probe. Measured values may be close to predicted values, as shown in Figure 3.16, but testing also may reveal errors or further complexities of the window, as shown in Figure 3.17,

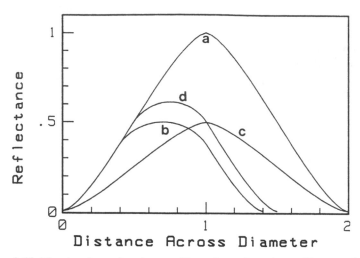

Figure 3.15 *The signals predicted as rectilinear bars of varying width (w) and reflectance (r) pass across a circular optical window with a diameter = 1. The position on the window diameter reached by the leading edge of the bar is shown on the x-axis. A bar with w = 1 and r = 1 is shown by line a, with w = .4 and r = 1 by line b, with w = 1 and r = .5 by line c, and with w = .5 and r = 1 by line d (Swatland, 1993b).*

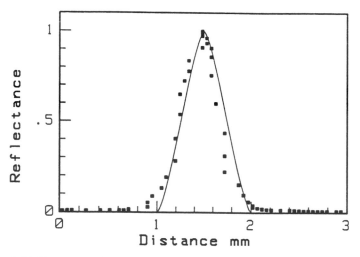

Figure 3.16 *Experimental detection of a white test bar (r = 1, w = .5 mm) against a black background by a single optical fiber (d = .5) in a light guide window, with illumination of the test bar from other fibers in the light guide. Predicted and measured signals are shown by a solid line and squares, respectively (Swatland, 1993b).*

72

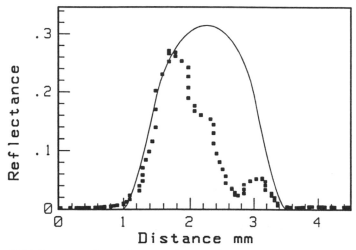

Figure 3.17 *Detection of a test bar (r = 1, w = .5 mm) by half of the fibers of a light guide group (d = 2 mm), with illumination by the remaining fibers. Predicted and measured signals are shown by a solid line and squares, respectively (Swatland, 1993b).*

where the difference between predicted and actual signals was caused by the nonrandom arrangement of optical fibers in a light guide. Half of the fibers were used to illuminate the bar, while the remainder were used to monitor its position.

Figure 3.18 illustrates a window problem originating from the curvature of an optical fiber to form a side window on a probe shaft. Placing the elliptical window close to the curvature of the optical fiber within the probe shaft increases the light intensity at the outer radius of the curve, so that light radiates from the optical window asymmetrically. Thus, in Figure 3.18, the light radiated forward in advance of the window and detected the bar before it actually reached the window. Thus, when testing a probe window, it is useful to measure the light intensity radiating from the window by moving a photometer in an arc around the window. In Figure 3.19, for example, the asymmetry of the optical window that produced the effect shown in Figure 3.18 may be seen.

One of the important consequences of the vignette effect in understanding signals from on-line meat probes is the relationship between the size of an anatomical structure and the base width of the peak it creates in the probe signal. If an anatomical structure has a narrow width relative to the diameter of the window, the base width of its peak is a little larger than the diameter of the window. For example, measured in relative units of window radii, with a window diameter, $d = 2$, and a bar width, $w = $

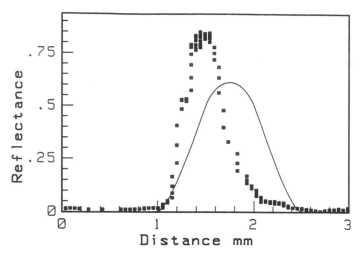

Figure 3.18 *Detection of a test bar (ᵣ = 1, w = .5 mm) by a single, large-diameter optical fiber (d = 1 mm), with light passing in both directions in the same fiber. Predicted and measured signals are shown by a solid line and squares, respectively (Swatland, 1993b).*

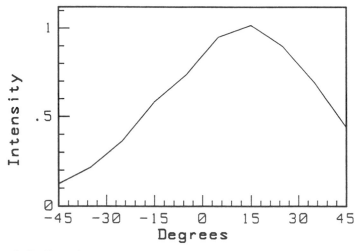

Figure 3.19 *Goniophotometry of light radiating from an optical fiber on a meat probe. The 0° position shows light radiating perpendicular to the shaft of the probe, while negative degrees are towards the handle of the probe and positive degrees are towards the tip of the probe (Swatland, 1993b).*

74

0.2, then the base width is $d + w = 2.2$, as shown for line a in Figure 3.20, and the peak intensity of the signal never reaches the maximum value that would occur if the structure completely fills the window. When the width of the anatomical structure equals the diameter of the window (line b in Figure 3.20), the base width of the signal is twice the diameter of the window but still is given by $d + w = 2 + 2 = 4$. The relationship also holds for structures that are large relative to the window, as for the case of a bar with a width of four radii ($d + w = 2 + 4 = 6$, line c in Figure 3.20).

Spectral Window Effect

Optical fibers have many applications in the on-line evaluation of meat quality because they enable a variety of optical measurements to be made directly on biological tissues, as clearly envisaged by the pioneers in the field (Kapany, 1967). However, when optical fibers are interfaced directly with meat, it is difficult to predict the effects of wavelength on the optical properties of the window and in the tissues beyond. The refractive index of the fluid phase of a biological sample may affect the Fresnel reflectance losses at the optical window, which are comparable in principle to those that occur in the coupling of optical fibers used for communication (Hewlett-Packard, 1982). When passing from one

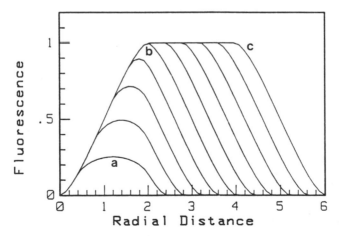

Figure 3.20 *Test bars of different width (0.2 to 2 window radii in width) but equal fluorescence (fluorescence = 1) pass an optical window to create peaks in the probe signal.*

medium (x) to another (y), with refractive indices n_x and n_y, respectively, then,

$$\text{Fresnel loss (dB)} = 10 \log ((2 + (n_x/n_y) + (n_y/n_x))/4)$$

Since both the refractive index and transmittance of optical fibers are related to wavelength, it is difficult to predict how white light will be transmitted into meat juices varying in refractive index.

When optical fibers are used for communication, generally, they are cut perpendicularly to the long axis of the fiber, unless a special type of coupling is in use. But a perpendicular end window is unsuitable for most types of meat probe. For example, if an optical fiber is used in a meat probe, the probe will require a cutting edge at its tip and the best location for the optical window is to one side of the tip or on the side of the shaft. Given that a narrow shaft on a meat probe has many advantages over a thick one and that optical fibers have a limited radius of curvature, it may be necessary to terminate the fibers at an acute angle at the optical window, rather than perpendicularly. This adds another dimension to the problem of understanding the optical window.

Apparatus such as that shown in Figure 3.21 allows reflectance at the internal surface of the window on a meat probe to be examined. The logic of the system is that, since it is difficult to get into the meat and measure the light coming into the meat through the window, equivalent information may be obtained by working from the other side of the window, getting into the optical system to see which wavelengths fail to pass into the meat because they are internally reflected at the window, versus those wavelengths that pass through the window into the meat.

If the distal window of an optical fiber is cut at 90° to the longitudinal axis, standardized to give a reflectance of 1 in air ($n = 1$), and then placed into distilled water, there is a decrease in internally reflected light (from a to b in Figure 3.22). The large change in internal reflectance from air to water shows why standardization of an optical fiber system from a white standard in air may be misleading when measurements are to be made with the optical fiber pushed into meat fluids. But the differences between water ($n = 1.333$) and 63.5% sucrose ($n = 1.450$) are far less, so that small differences in the refractive index of meat fluids were not important for the plastic optical fiber on which these measurements were made (Swatland, 1991c). However, the shape and magnitude of this effect depends on the type of optical fiber and should be checked for a window likely to be exposed to meat fluids differing in refractive index (Chapter 5).

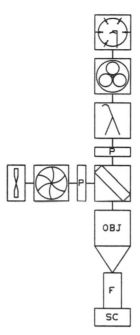

Figure 3.21 *Apparatus for examining internal reflectance at the window of an optical meat probe. Crossed polars (P) remove reflectance at the top or distal window of the optical fiber (F), so that a microscope objective lens (OBJ) may be used to couple the optical system to the distal window of the optical fiber, which is inserted into a sample chamber (SC) containing fluids of different refractive index (Swatland, 1991c).*

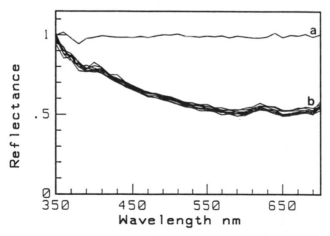

Figure 3.22 *Internal reflectance from the distal window (90°) of an optical fiber standardized in air (a) and then inserted into fluid (b) ranging from water (n = 1.333) to approximately 63.5% sucrose (n = 1.450).*

The angle at which the distal window of an optical fiber is cut (relative to the longitudinal axis of the optical fiber) also may have a strong effect on internal reflectance, as shown in Figure 3.23, where a fiber cut at 90° was set at a reflectance of 1. Shaving wedges off the perpendicular endface first reduced the internal reflectance, then increased it. In other words, cutting this optical fiber at an angle of about 75° to form an elliptical window increased the amount of light that passed into the meat.

Probe Velocity

One of the major factors affecting the performance of a meat probe is the velocity of penetration, which may interact with the response time and saturation characteristics of the photometer. For a handheld probe, it is essential to ensure that velocities that are comfortable and sustainable for the operator fall within the range of optimum performance of the instrument. For example, a back-fat probe that works satisfactorily when used at a normal, relatively low velocity may fail when used at a very high velocity by an angry or frustrated operator. Not only must we specify the acceptable velocity range, especially if the probe is used in a statutory meat grading program, but, also, the probe must have an error light to alert the operator to a velocity-related failure. Referring back to Figure 1.4, it is evident that, for a probe that is triggered by an internal clock to make measurements at set time intervals, a high velocity will

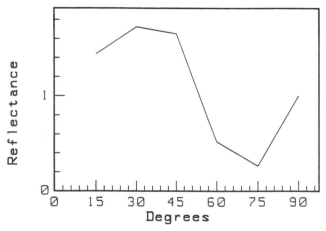

Figure 3.23 *Internal reflectance from the distal window of an optical fiber cut at various angles to the longitudinal axis of the fiber.*

result in too few data being collected to find the cutoff. With just a few data points between F_{max} and M_{min} on Figure 1.4, the cutoff is likely to be positively biased (incorrectly located at too great a depth). Conversely, if the probe is used at a very slow velocity, the large number of data collected may exceed the storage capacity or take an excessive time to process. Alternatively, for a probe that is triggered to make measurements at set increments of depth, a very high velocity may force the probe to make measurements at a rate faster than that allowed by the photometer response time or the analogue to digital converter for the optical signal. In Figure 3.5, for example, measurements at fast shutter speeds underestimate the full intensity of the light source.

When examining bidirectional velocity effects (using both way-in and way-out data), it is important to ensure that measurements are compatible. For example, if a computer, rather than dedicated hardware, is used to operate a probe, it may be difficult to make two measurements (depth and reflectance) at exactly the same time (because the computer executes instructions sequentially). For example, if the depth is measured after reflectance on the way-in, the elasticity of the meat will cause a hysteresis effect when the way-in measurements are compared with the way-out measurements. To avoid this, the software should reverse the order of measurement when the probe changes direction at the maximum depth in the meat.

Figure 3.24 shows three positions on an ideal signal used to study the effect of probe velocity on signal deformation (at peak height, *a;* at half peak height, *b;* and at the base, *c*). The ideal signal was created by moving a test bar slowly past the meat probe window. Unfortunately, the dip in the top of the peak was from a defect in the test bar, apparently created during manufacturing of the paper (variation in paper thickness) or by ink smearing in the printer. The slight slopes in the sides of the peak are explained by the relationships shown in Figure 3.20, coupled with a slight photometer delay to give deviations in the ideal signal of $a = -8.9\%$, $b = 5.5\%$, and $c = 23.5\%$. Thus, the peak started when the test bar reached the window and did not finish until the bar had completely passed the window. This accounts for the positive deviation at measurement *c* at the base width (Figure 3.24). Conversely, maximum fluorescence was not obtained until the test bar fully blocked the window and was only maintained while the bar blocked the window, thus accounting for the negative deviation of measurement *a*. In Figure 3.24, the way-in data are shown by a solid line, and the way-out data are shown as solid squares. Measurement *d* was the *x*-axis separation of way-in from way-out data at half peak-height ($d \approx 0$ where way-in and way-out traces are almost identical).

Figure 3.25 shows the distortion of the signal caused by probe velocity. It should be emphasized that this was a laboratory prototype, and serious problems would be expected industrially for a commercial probe with these relatively poor performance characteristics. Way-in and way-out measurements were averaged for measurements a, b, and c. Measurement b was least affected by probe velocity, whereas measurements a and b were unacceptable at high speed. Measurement d increased at higher velocities, probably because of hysteresis from tissue elasticity. The bias was negative for way-in depth readings and positive for way-out. This was corrected by damping the fluorescence signal at 10 Hz, thus slowing the rate of change of the photometer output to match that of the depth sensor. With this prototype at velocities <5 cm sec^{-1}, correlations of test bar fluorescence with peak signal intensity were typically $r = -0.96$. At higher speeds, there was insufficient time to reach full intensity on the narrowest bands, although the effect was not serious.

Attempting to cut through the details to reach the main point, it is

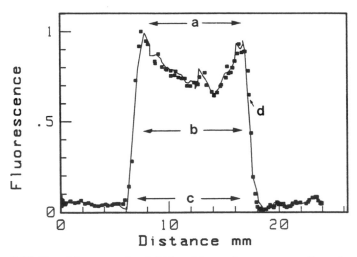

Figure 3.24 *Signal from a test bar (width = 10 mm, fluorescence = 1) moving slowly (≈ 10 mm sec^{-1}) past an elliptical optical window (maximum diameter 1.1 mm) on a connective tissue probe. Way-in measurements are shown as a line and way-out measurements as solid squares. At higher velocities, distortions from this ideal signal were measured as the deviation percent from 10 mm ideal width at (a) maximum peak height, (b) half-peak height, and (c) the base. The x-axis separation of way-in and way-out traces was measured at half-peak height (d $\approx$ 0 in this ideal signal).*

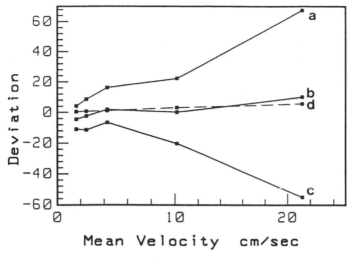

Figure 3.25 *Effect of probe velocity on percent deviation of measurements* a,b, *and* c, *and on millimeters deviation of measurement* d *(Swatland, 1992).*

evident from Figure 3.15 that it is possible to correlate a probe signal peak with the anatomical structure that created it, but peak width may be a complex function of anatomical width. Similarly, peak intensity may be a function of both the optical properties of the anatomical structure and its width. There are geometrical reasons for using the base width of an optical peak to assess the size of anatomical structure detected by the probe, but the width at half-peak height may be less susceptible to velocity-related errors.

STANDARDIZATION

Standardization is essential for making scientific measurements and highly desirable for objective measurements in the on-line evaluation of meat. The most difficult situation is measuring absolute values (as in fluorometry of connective tissue) because the problems involved are quite formidable. Even using a radiometer calibrated by the manufacturer against a national standard, it is virtually impossible to set and maintain the excitation at a standard intensity and equally difficult to standardize the measuring conditions. Easier situations arise when a ratio may be used, such as reflectance, transmittance, or absorbance, which reduces the problem to measuring light in versus light out. But, even here, a standard is required that can be called 100%.

Both the apparatus and the standardization protocol have important effects on measurements made with optical meat probes. Figure 3.26 illustrates the changes produced by modifying the probe tip of an otherwise unchanged xenon-flash and diode-array spectrograph system, as shown in Figure 3.11. Spectrum 1 was obtained with illuminating and receiving optical fibers forming a scattered pattern in the common trunk of a bifurcated light guide. The probe was placed in contact with the cut surface of the pork, and the general shape of the spectrum is similar to those obtained by macroscopic reflectance spectrophotometry (Chapter 7). Spectrum 2 in Figure 3.26 shows results obtained from pork with a different probe, but the same light source and spectrograph. Illuminating and receiving fibers radiated to form an annular window on the lateral surface of a spear-type probe inserted deep into the meat. For spectrum 2, the apparatus was standardized with white tape (polytetrafluoroethylene, PTFE) wrapped around the window (ten layers deep without stretching). Thus, changing the design of an optical window and the way in which it is used may produce major changes in the result.

For standardization, the basic problem is the difficulty of finding a

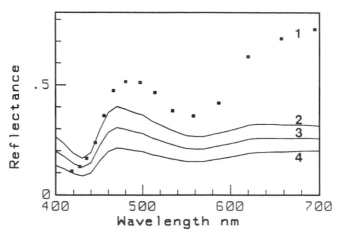

Figure 3.26 *Effects of window geometry and method of standardization on measurements of pork made with a xenon-flash and photodiode array spectrograph. Spectrum 1 was obtained with parallel illuminating and receiving optical fibers butting onto the meat surface, while spectra 2, 3, and 4 were obtained with the illuminating and receiving optical fibers in a radial pattern to produce an annular window on a cylindrical shaft. Spectrum 2 was standardized against PTFE tape, spectrum 3 against the mean intensity of vibrating barium sulfate powder, and spectrum 4 against the maximum intensity of vibrating barium sulfate powder.*

white standard that can be molded to a window with a complex shape. For a simple window, as in the flat end of the common trunk of a bifurcated light guide, a plate of optical quality barium sulfate may be newly pressed and placed parallel to the window. By adjusting the distance between the window and the powder surface (taking care not to touch the powder surface), a distance may be found that gives the maximum photometer response; then, reflectance is set at 1 for all wavelengths. The system should be tested by remeasuring the standard at several different heights, all of which should give a series of flat spectra (at different intensities). Sometimes, the spectra are not flat if there is a protective layer over the probe window. Some type of protection is highly desirable, however; otherwise, meat juices penetrate the small channels between the optical fibers in the light guide and slowly change its performance. For everyday use, the plate of barium sulfate may be replaced by a secondary standard of opal glass, always remembering the appropriate corrections that must be made to the data.

The protocol described above cannot readily be used if the optical fibers open in a ring around the side of the probe shaft. The probe might be placed in a tube with its internal surface coated with magnesium oxide, but the levels of light returned to the spectrograph are relatively low compared with other methods. Spectra 2, 3, and 4 in Figure 3.26 show differences in the reflectance spectrum of the same sample of pork caused by the method of standardization. For spectrum 2, the ring of optical fibers was standardized against PTFE tape wrapped around the shaft. Spectrum 3 was standardized by placing the probe into a beaker full of optical-quality barium sulfate powder vibrating at 60 Hz (stationary powder is no use because the probe presses against it to create a variable gap). Thus, as the amount of light returned to the spectrograph at standardization was increased with vibration of the barium sulfate, the same sample of pork appeared to have a lower reflectance than when the probe was standardized against PTFE tape. Even with vibrating powder, there are chance events that lead to variation in the intensity of light returned to the spectrograph. Using the maximum light intensity of vibrating barium sulfate for standardization, as in spectrum 4, makes it appear that the pork sample has an even lower reflectance than when standardized against the mean for vibrating powder.

Thus, relative to conventional surface reflectance spectrophotometry standardized against magnesium oxide, it is extremely difficult to standardize meat probes. If we were to choose one arbitrary standard for us all to use, it would be PTFE tape, although care must be taken to avoid stretching and to control the number and thickness of layers of tape. The optical properties of PTFE are described by Weidner and Hsia (1981).

REFERENCES

Andersen J. R., C. Borggaard, T. Nielsen, and P. A. Barton-Gade. 1993. *39th International Congress of Meat Science and Technology,* Calgary, Alberta, pp. 153–164.

Aunan, W. J. and L. M. Winters, 1952. *J. Anim. Sci.,* 11:319.

Birth, G. S., C. E. Davis, and W. E. Townsend. 1978. *J. Anim. Sci.,* 46:639.

Conway, J. M., K. H. Norris and C. E. Bodwell. 1984. *Amer. J. Clin. Nutr.,* 40:1123.

Eichinger, H. M. and G. A. Beck. 1992. *Archiv Tierzucht,* 35:41.

Hazel, L. N. and E. A. Kline. 1952. *J. Anim. Sci.,* 11:313.

Hazel, L. N. and E. A. Kline. 1959. *J. Anim. Sci.,* 18:815.

Hewlett-Packard (1982). *Digital Data Transmission with the HP Fiber Optic System.* Application Note 1000. Palo Alto, CA: Hewlett-Packard.

Hildrum, K. I., B. N. Nilsen, M. Mielnik, and T. Naes. 1994. *Meat Sci.,* 38:67.

Kapany, N. S. 1967. *Fiber Optics. Principles and Applications.* New York: Academic Press.

MacDougall, D. B. 1980. *J. Sci. Food Agric.,* 31:1371.

Norris, K. H. 1984. Reflectance spectroscopy, in *Modern Methods of Food Analysis,* K. K. Stewark and J. R. Whitaker, eds., Westport, CT: AVI Publishing Co.

Swatland, H. J. 1982. *J. Food Sci.,* 47:1940.

Swatland, H. J. 1986. *Can. Inst. Food Sci. Technol. J.,* 19:170.

Swatland, H. J. 1991a. *J. Anim. Sci.,* 69:1983.

Swatland, H. J. 1991b. *Comput. Electron. Agric.,* 6:225.

Swatland, H. J. 1991c. *J. Comput. Assist. Microsc.,* 3:233.

Swatland, H. J. 1992. *Comput. Electron. Agric.,* 7:285.

Swatland, H. J. 1993a. *J. Comput. Assist. Microsc.,* 5:231.

Swatland, H. J. 1993b. *J. Anim. Sci.,* 71:2666.

Thodberg, H. H. 1993. *Neural Comput. Applic.,* 1:248.

Tsuruga, T., T. Ito, M. Kanda, S. Niwa, T. Kitazaki, T. Okugawa, and S. Hatao. 1994. *Meat Sci.,* 36:423.

Weidner, V. R. and J. J. Hsia. 1981. *J. Opt. Soc. Amer.,* 71:856.

Workman, J. and H. Andren. 1992. *Amer. Lab.,* 24(12):20R.

CHAPTER 4

Electromechanical Properties of Meat

INTRODUCTION

SHAKING HANDS WITH the pig may have been an old custom for country butchers. After a pork carcass has been dressed and is hanging from a gambrel, its forelimbs project outwards, away from the jowl and belly, and shaking the forelimb in a fore and aft walking motion may enable the butcher to detect the early onset of rigor mortis in a PSE carcass. Later on, the same shaking motion gives information about setting of the fat, which makes the meat stiff enough to cut neatly. Eventually, shaking hands may convey information about softening of the meat as it is conditioned, becoming more tender, succulent, and tasty. The play on words regarding shaking hands may be unfamiliar, but the cut of meat called a picnic in North America used to be called a hand and spring of pork. Another mystical skill, only learned by practical experience, is how to judge softness of PSE pork. Pushing the tip of the index finger into the horizontal surface of a cut longissimus dorsi on a recently ribbed hanging carcass may reveal much about the softness of the meat and its future fluid losses. Sometimes the meat springs back when the fingertip is removed, sometimes it comes back slowly, and sometimes a depression remains and fills with fluid. Similarly, pinching a beefsteak tightly between thumb and forefinger may convey a good impression of its future eating characteristics.

These examples indicate that the skilled butcher is using touch, aided by vision, to glean subtle information about intrinsic properties of meat, which is the same objective we have in the on-line evaluation of meat. Subjective touch translates into objective rheology and kinetics, and this chapter gathers together a number of topics that all involve the mechanical properties of meat. The range of topics is from the ponderous kinetics of the whole hanging carcass through to the transmittance of ultrasound through the microstructure of meat, with side excursions to consider on-line monitoring of slaughter operation, the mechanical performance of meat probes, and robotic delivery systems for probes.

ON-LINE MONITORING OF SLAUGHTER OPERATIONS

Electrical Stunning

Electrical stunning is used extensively for pigs, sheep, calves, and poultry. Monitoring electrical stunning on-line is relatively simple (compared with monitoring carbon dioxide or concussion stunning), and this enables the performance of the stunning system to be checked continuously. Thus, problems with animal loading, corroded electrodes, and inadequate washing of animals may be detected almost instantly. With our increasing concern for the humane treatment of meat animals, it is essential that every animal is humanely stunned prior to slaughter (except for approved methods of ritual slaughter where the religious rights of potential voters seem to outweigh our obligations to treat nonvoting farm animals in a humane way). If the burden of legal proof is placed on the slaughterer, it would be possible to archive digital stunning data inexpensively on tape for future scrutiny.

Published data on commercial stunning systems are scarce, so Figure 4.1 shows the stunning of a pig in a research pilot plant (the duration of stunning is longer than in a commercial abattoir). There is extensive literature on electrical stunning for pigs, the general consensus of which is that 300 V AC, with a short stunning time and a minimum delay between stunning and sticking, causes the least amount of PSE (van der Wal, 1978). In Figure 4.1, V-shape electrodes were in place on the head of the pig before the operator switched on the stunning current. The current was approximately 1 A.

Surgically implanted recording electrodes on the pig's brain have been used to find the voltage that reaches the brain. It is far less than the voltage applied externally because of the relatively high resistance of hair, skin, and bone sinuses (Figure 4.2). Electrocorticography, recording from surgically implanted electrodes on the brain, is an excellent method of studying brain activity because it bypasses the strong electromyographic activity of jaw and eye muscles (Swatland, 1976b), but it would be impossible on-line. Electrocorticography is a superior method of detecting localized brain activity, which may be used to assess the degree of animal consciousness. However, electroen-

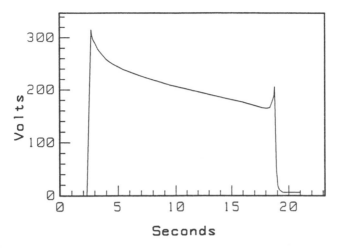

Figure 4.1 *On-line monitoring of the interelectrode voltage applied externally during the stunning of a pig.*

cephalography, using surface electrodes on the head, may be informative under experimental situations (Newhook and Blackmore, 1982a, 1982b; Blackmore and Newhook, 1982) and is more feasible as an on-line method. The major problems in developing an on-line monitoring system are interference from the electrical activity of jaw muscles and the difficulty of automatic placement of the electrodes,

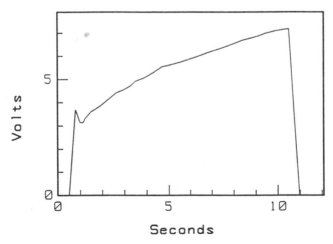

Figure 4.2 *Monitoring of the voltage across the pig's brain from surgically implanted electrocorticography electrodes.*

which have far more stringent operating requirements than stunning electrodes.

The decrease in applied root mean square voltage seen in Figure 4.1, during which amperage was almost constant, may have originated from venous pooling in the sinuses around the brain, thus lowering resistance. Figure 4.2 shows that the voltage reaching the brain increased with time. Massive changes in blood distribution occur during stunning, since most of the musculature contracts in a seizure. This may cause blood spots or petechial hemorrhages in the meat (Gilbert and Devine, 1982), which sporadically become a serious commercial problem. On-line monitoring of stunning currents could be useful in finding the causes. There are many other topics of importance in this area, such as the survival time of neural pathways from the central nervous system to the musculature (Swatland, 1975, 1976b), but they are only indirectly related to on-line evaluation.

In summary, on-line monitoring of animal stunning might be required in the future to comply with animal welfare legislation. Information may be gained by monitoring the stunning current, detecting, and immediately correcting improper electrode contact with the animal. Additional information might be gained by electroencephalography, recording through the electrodes just used for stunning. The major problem in pigs is jaw muscle activity; however, if an animal is delayed to monitor its response to stunning, this prevents immediate exsanguination after stunning.

Reflex Muscle Activity

Reflex activity of muscles just after stunning and exsanguination is a major safety concern for personnel, but it also has long-term effects on muscle glycolysis (Bendall, 1973), thus affecting pH and all the pH-related properties of meat. Muscle reflex activity may be monitored by electromyography using surface electrodes, as shown in Figure 4.3, which shows muscle activity during exsanguination of a pig (Swatland, 1976a). Surface electrodes are impractical for routine use because either the wires get tangled or the telemetry transmitter gets detached.

Contraction of individual muscle fibers is imperceptible in a whole carcass, and visible movements of large parts of the body require neural coordination. Loss of excitability in the axons of motor neurons follows a proximal to distal sequence preceded by hyperexcitability, and the rate of loss of excitability depends on temperature.

Although seemingly trivial events at the end of an animal's life, and

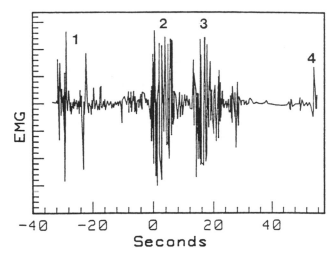

Figure 4.3 *Electromyography during exsanguination of a pig at zero seconds. The pig had a moderate amount of muscle reflex activity after shackling (1), which was followed by two major bursts of muscle activity shortly after exsanguination (2 and 3), and then the carcass was still except for occasional movements (4).*

thus ignored by physiologists, the sources of reflex activity in carcasses are extremely complex. Motor neurons in the ventral horn of the spinal cord receive both inhibitory and excitatory inputs, but the descending and ascending tracts that link the brain and spinal cord in meat animals are polysynaptic, so that motor neurons in cattle, sheep, and pigs have a greater degree of autonomy than in humans. Meat animals have only a small direct motor cortex in the cerebrum and only short motor tracts descending the spinal cord, and the sensory areas in the cerebral cortex are dominated by inputs from the lips and snout, while areas corresponding to the limbs and trunk are very small.

Neural activity may originate from motor neurons released from their normal inhibitory control. Well-defined physiological reflexes, such as the myotactic or stretch reflex of muscles, only occur briefly for a short while after exsanguination because they involve a sensory input from neuromuscular spindles and a sensitive spinal reflex to resist a stretching force imposed on the muscle. The reflexes that have the most effect on muscle glycolysis are crude, extensive contractions that activate large muscles, just as if they had been electrically stimulated. The summation of these events may be seen by monitoring the output of a load cell in the shackling chain (Swatland, 1983). Not only is it relatively simple to integrate this activity, obtaining a single index for each animal, but it may be possible to distinguish which parts of the carcass are involved (Figure 4.4). On-line monitoring of reflex activity may reveal whether stunning and exsanguination are being done properly. Extensive reflex activity of the type shown in Figure 4.4 is highly likely to make any pork carcass become PSE. A load cell in the shackling chain also may be used to monitor exsanguination (Figure 4.5).

In practical terms, operations such as the electrical stunning of pigs are relatively simple to monitor on-line. The combination of stunner current and shackling chain load cell gives some useful information for demonstrating compliance with humane slaughter standards, as well as providing forewarning of carcasses that are likely to become PSE because of excessive reflex activity. In a population without any PSS (PSS pigs may become PSE without excessive reflex activity), this information is early enough to enable PSE carcasses to be streamed towards an accelerated cooling line involving a hide-puller, rather than scalding and dehairing, followed by accelerated refrigeration.

Electrical Stimulation Monitor

Electrical stimulation of carcasses soon after slaughter accelerates the decline in pH and the development of rigor mortis, thus minimizing the possibility of cold-shortening. Also, there may be beneficial effects on meat tenderness and improvements in the USDA beef grading scores for

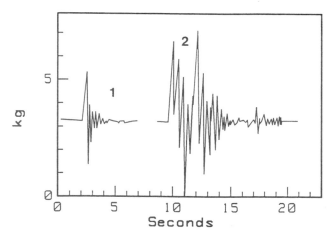

Figure 4.4 *Major reflex activity of the unshackled ham (1) and both forelimbs (2) in a pork carcass detected on-line with a load cell in the shackling chain.*

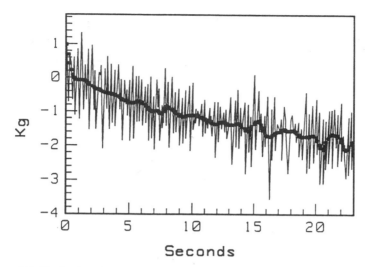

Figure 4.5 *Monitoring the exsanguination of a pig from a load cell in the shackling chain. The bouncing is caused by reflex activity, while the integrated signal (thick line) shows the weight loss.*

youthfulness and meat color (Savell et al., 1978), although cooking losses may be increased (Bouton et al., 1980). The strength and duration of stimulation required to produce these effects depend on how long after slaughter the stimulation is applied. Immediately after slaughter, when the nervous system is still functional, weak stimulation may be amplified by the nervous system to produce an overwhelming response, sometimes too excessive. Later on, after neural activity has declined, high voltages may be required to produce even moderate contraction. There are other factors involved, such as electrode placement and effective area, stimulation waveform and frequency, and differences in intrinsic muscle physiology. Thus, knowing whether or not a muscle has been adequately stimulated may be difficult. In New Zealand, lambs are stimulated at 100 V for a minimum of ninety seconds, and a clip-on monitor has been developed to record stimulation time, peak current, pulse width, and frequency (Loeffen, 1992).

ON-LINE MUSCLE RIGOROMETRY

The major change that marks the conversion of muscle to meat is rigor mortis, when filament sliding no longer is possible and sarcomere lengths become fixed. This occurs in muscles that have depleted their energy reserves or have lost the ability to utilize their remaining reserves. A myosin molecule head on a thick filament, attached to an actin molecule on an adjacent thin filament, can only detach itself if a new ATP molecule is available. When muscle is converted to meat, myosin molecule heads become permanently attached to actin molecules so that contraction and passive filament sliding or stretching become impossible.

Information Content of Rigorometry

Loss of muscle extensibility may be measured with a rigorometer (Bate Smith, 1939; Bendall, 1973). In a laboratory rigorometer, the lower end of a hanging muscle strip is gently loaded, and the resulting elongation of the muscle is recorded. When a muscle strip still containing ATP is loaded, the lower end of the muscle drops as the muscle stretches, and, when the load is removed, the muscle returns to its original length. But when ATP no longer is available, because energy reserves have been depleted or enzymes have been inactivated, the muscle is only very slightly extensible. Muscle strips normally are maintained in an anaerobic atmosphere to prevent surface resynthesis of ATP, which is far more efficient aerobically than anaerobically. The general idea is to simulate the conditions occurring anaerobically within the bulk of the muscle mass of the carcass.

The extent to which a muscle strip stretches when loaded depends on its initial length and cross-sectional area and on the load applied. Elongation of an elastic body is proportional to the applied force, until the body reaches its limit of elasticity (Hooke's Law). Muscle properties are similar, but the internal structure is much more complex, since the muscle contains a variety of mechanical components in series and in parallel, all surrounded by fluids with a variable viscosity and microstructural flow path. Young's modulus (Y) is used to make allowances for differences in sample dimensions and load.

$$Y = \frac{k\,dL}{A} \times \frac{F/A}{dL/L}$$

where A is the sample cross-sectional area and L is the initial sample length. As rigor mortis develops in beef muscle, the modulus increases from about 1110 to 34,900 (Marsh, 1954). Thus, with a constant load and an increasing modulus, the lengthening of the muscle strip is diminished.

ATP may be resynthesized from creatine phosphate for a variable length of time after slaughter, and, when it is all depleted, the length of time before the occurrence of rigor mortis depends on the amount of glycogen available within the muscle and on the survival of glycolytic enzymes. This creates a *delay* period of constant elongation when the muscle is loaded because ATP is still being resynthesized, a *rapid phase* of change in muscle lengthening as individual muscle fibers run out of ATP, and a final *postrigor phase* when the majority of fibers resists lengthening to prevent any further extensibility of the whole muscle strip. The terms in italics are the standard terminology (Bendall, 1973). Sometimes, muscle strips contract weakly as they develop rigor mortis, which complicates further measurements of extensibility.

The response of a muscle strip in a rigorometer is determined by the condition of the animal at the time of slaughter (Bendall, 1973). Animals that are calm during slaughter produce muscle strips with a long delay, a slow rapid phase, and a decrease in unloaded length at body temperature but not at room temperature. Animals that struggle during slaughter produce a short delay, a short rapid phase, and a decrease in unloaded length at body temperature but not at room temperature. Exhausted animals produce an extremely short or nonexistent delay period and a very short rapid phase, and muscle strips may shorten at both body and room temperature. Starved animals produce a short delay period, a relatively long rapid phase, and a decrease in unloaded length at body temperature but not at room temperature.

Shortening of muscles as they develop rigor mortis also depends on muscle temperature, since superficial muscles cool faster if they are not extensively insulated by fat. Some muscles are completely prevented from shortening because they are already stretched by the weight of the hanging carcass; however, unrestrained muscles along the loin and posterior hind-limb of a hanging carcass are free to contract. In PSE pork carcasses, rigor contraction may be sufficient to lift the forelimb of the hanging carcass (Davis et al., 1978).

Unrestrained muscles that contract as they develop rigor produce meat with short sarcomeres, a greater overlap of thick and thin filaments, and a greater number of rigor bonds linking the filaments. Thus, shortened muscles cause commercial problems because they are tough and have a decreased water-binding capacity. The extent of rigor contraction of unrestrained muscles may be reduced by hanging carcasses from the pelvis, but then many retail cuts of meat lose their familiar appearance, and the normal process of tenderization caused by muscle stretching is lost in other muscles.

Rigorometry would be a useful technique to apply on-line in the meat industry, but the direct measuring approach used in the laboratory rigorometer is clearly unrealistic because it involves removing a strip of muscle; however, one of the major criticisms of laboratory rigorometry is that it requires the removal of a muscle strip from the carcass. No matter how carefully this is done, it involves cutting axons, electrically shorting transected membranes, and altering the gaseous environment and temperature for a while. Removal of the sample itself, therefore, is a major trigger to glycolysis that does not occur in the muscle mass that the strip is supposed to emulate.

Surface Deformation On-line Rigorometer

The first on-line rigorometer avoiding these problems was developed by Sybesma (1966). It was portable and spring-loaded and was pushed into the surface of carcass muscles to make a measurement (Figure 4.6). Sybesma (1966) measured over 1100 hams at forty minutes postmortem and found that 31 % had already set in rigor mortis. Variation was greater in the semimembranosus than in the adductor (Figure 4.7), and there was tentative evidence of a bimodal distribution in semimembranosus, which might have been caused by PSS pigs (compatible with the tentative bimodality seen in Figure 2.1). One would expect PSE pork from PSS pigs to develop rigor mortis more quickly than normal pork. Thus, if searching for this effect, semimembranosus might be more useful than adductor. In semimembranosus, rigor development was correlated with both pH and temperature, $R = 0.59$.

Tensile Rigorometer On-line

Other on-line rigorometers have been tested, but they are less robust than the simple handheld rigorometer developed by Sybesma (1966), although they allow continuous monitoring of experimental carcasses in a pilot plant. Rather than pushing on the muscle, which involves connective tissue deformation and stiffening of fat from refrigeration, an alternative method is to pull on an exposed muscle fasciculus. Both force and distance may be monitored to obtain superior resolution of the rheological properties of the meat as it develops rigor mortis. In the

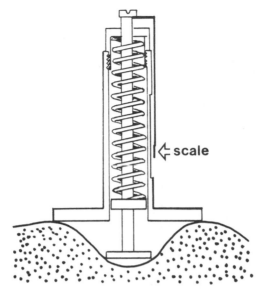

Figure 4.6 *Sybesma's on-line rigorometer (Sybesma, 1966).*

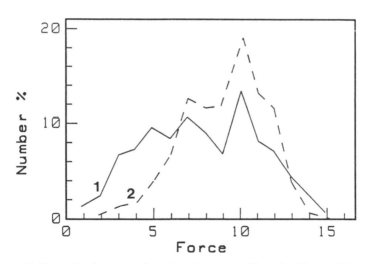

Figure 4.7 *Rigor development of semimembranosus (1) and adductor (2) muscles measured on-line in over 1100 hams by Sybesma (1966).*

rigorometer shown in Figure 4.8, the base plate of the apparatus is attached to the carcass by a screw into the pubis (1) and to the adductor muscle by several loops of surgical thread (2). An arm (3) swivels around a pivot (4) when pushed by a solenoid (5), thus pulling on the muscle. The force exerted on the muscle is read from a strain gauge (6), and the distance moved is read from a linearly variable differential transformer (7).

A typical result is shown in Figure 4.9. The method is not much use for studying rigor development in PSE carcasses because most severely affected carcasses are in rigor before the rigorometer can be attached. However, in normal pork carcasses with slow rigor development (two to three hours postmortem), the method gives reliable results that are strongly correlated with other indices of rigor development, such as the loss of membrane capacitance and loss of ATP measured by nuclear magnetic resonance.

Pneumatic Rigorometer On-line

The tensile rigorometer only allows superficial muscles of the carcass to be tested, whereas the pneumatic rigorometer allows deeper muscles to be tested. In the controller shown in Figure 4.10, an air pump (1) monitored by a pressure transducer (2) increases the air pressure over a

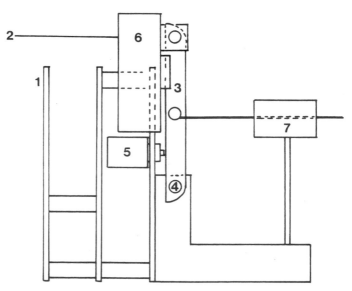

Figure 4.8 *Tensile rigorometer for use on a pork carcass (Swatland, 1985). Components are described in the text.*

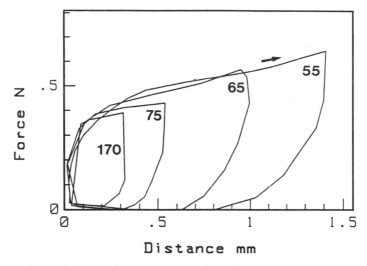

Figure 4.9 *Development of rigor mortis in the adductor muscle of a pork carcass hanging in a meat cooler (times in minutes). The arrow shows the direction round the hysteresis area.*

reservoir of saline solution (3), forcing it into another compartment (4) whose weight is monitored from a strain gauge (5). The resulting air pressure in the weighing cylinder is transmitted via narrow-bore tubing to the interior of the carcass where it causes the expansion of a soft bulb (6). Solenoid-operated valves provide connections to a source of saline solution (7) and to the atmosphere (8) for standardization and equilibration when the expansion bulb is inserted into the carcass. Saline solution is required to avoid freezing in the meat cooler. The pneumatic expansion bulb is contained within a sharp tube that is pushed deep in the muscles, the tube is withdrawn, and the skin is sutured around the narrow-bore tubing. Expansion of the bulb at regular intervals displaces muscles deep in the carcass, and their resistance is detected by the pressure-flow characteristics within the pneumatic controller.

A typical result is shown in Figure 4.11. The major problem with this method is that, sometimes, the connective tissues within the muscle split apart, allowing the bulb to expand into a larger space, thus losing the baseline reference. Also, the method is sensitive to the stiffening of fat, which occurs as the carcass is refrigerated.

Impact Rigorometry

Another possibility for the detection of rigor mortis is to use rapid

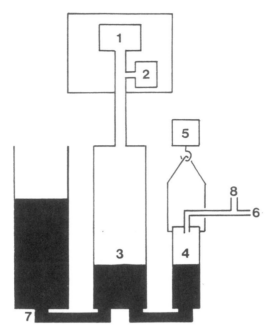

Figure 4.10 *Controller for a pneumatic rigorometer (Swatland, 1986). Components are described in the text.*

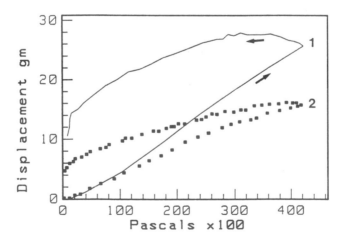

Figure 4.11 *Pressure flow characteristics of pork before (1) and after (2) development of rigor mortis measured with a pneumatic rigorometer (Swatland, 1986). The arrows show the direction round the hysteresis area.*

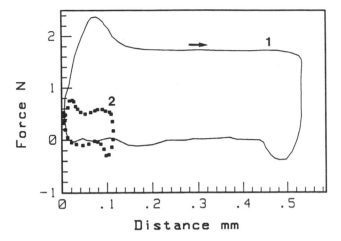

Figure 4.12 *Impact rigorometry of pork before (1) and after (2) development of rigor mortis. The arrows show the direction round the hysteresis area.*

compression, equivalent to slapping the meat by hand, which may give a good subjective impression of the degree of rigor development. A fast impact generated by a strong solenoid (120 V, 0.1 sec) gives adequate time to measure a force-deformation relationship, as in the example shown in Figure 4.12. Before the development of rigor mortis, the meat is very flexible, and the applied force is transmitted rapidly through meat, which then is deformed at a steady rate until the solenoid cuts out. The meat then pushes the solenoid back to its starting position, thus creating the negative forces seen in Figure 4.12. After rigor mortis has developed, the deformation of the meat is greatly reduced. The method has been shown to work well in the laboratory (Swatland, 1985) but has not yet been tested on-line. Combining this idea with that of the load cell in the shackling chain, it looks as if similar information could be generated on-line simply by reading from the load cell as the carcass passes down a small step in the overhead rail. Thus, the dropping of the carcass would provide the energy for deformation.

MEAT STRUCTURE AND TENDERNESS

Armour Tenderometer

The Armour Tenderometer (Hansen, 1972; Figure 4.13) can teach us much about the methodology of on-line meat evaluation: discovering

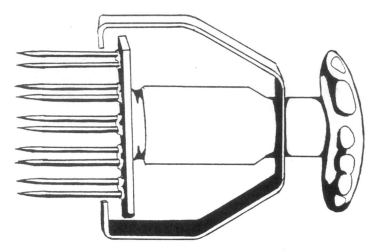

Figure 4.13 *Hansen's Armour Tenderometer (Hansen, 1972) built by BLH Electronics, Inc., Waltham, Massachusetts. The strain gauge between the operator handle and the needles was connected by cable to an instrument box displaying the results.*

why something did not work as well as expected is a way to discover how something can be made to work better. The Armour Tenderometer was developed in a logical manner, first examining the performance of penetration (needle), compression (disk), and shear (v-notch) probes on isolated meat samples using an Instron Universal Testing Machine, and then developing a probe design with the least variance. However, an arbitrary decision was made to use Warner-Bratzler shear values of meat roasted to an internal temperature of 65.6°C as a standard indicator of cooked tenderness, with a somewhat worrying assumption that, "Reliance on Warner-Bratzler shear force of cooked samples for a standard is more or less traditional and goes along with most published work on meat tenderness." For readers approaching the subject without previously having studied much meat science, this may not mean too much. But if one were to question meat scientists, two main groups might be distinguished: those who believe in Warner-Bratzler shear tests and those who do not. Everyone would agree it is faster, easier, and less expensive to use a shear test than a taste panel and that peer-reviewed scientific publications accept competent shear test measurements as sound data, but the doubters would argue that a strong correlation of shear test with taste panel is essential if laboratory results are to be extrapolated to real-world situations. Not only this, but it would be demanded that the correlation of shear test with taste panel should be established for a range of meat tenderness similar to that found commercially. Finding that old cows produce tougher beef than young steers may be a legitimate and useful way to investigate a phenomenon scientifically, such as the biochemistry of collagen development, but it tells us little we can apply directly in an industrial situation.

The final outcome is that, even if the Armour Tenderometer had been a complete technological success, perfectly correlated with the Warner-Bratzler shear test, still it would be limited in practical value by the strength of the correlation of shear test with taste panel. With luck, it might be found that the correlation of Armour Tenderometer with taste panel was even stronger than the correlation of Warner-Bratzler shear with taste panel. But, without luck, it might be found that the Armour Tenderometer had become sensitive to a source of experimental error in the electromechanical operation of the Warner-Bratzler shear apparatus.

The decision to use a battery of needles instead of a single needle also is instructive. Statistical analysis showed the standard error of the penetration force decreased sharply as the number of measurements was replicated to $n = 10$, with a lesser decline with further replication. However, each single penetration is an independent measurement, which hardly can be compared to the response of a single needle mounting with nine others on the same platform. But when the Armour Tenderometer was tested on U.S. choice beef, penetration resistance was correlated with taste panel scores, $r = 0.77$ (Figure 4.14). Unfortunately, this promising early result was not sustained in further testing (Table

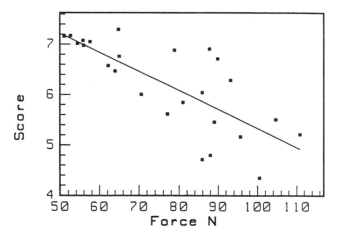

Figure 4.14 *Penetration resistance (force) of U.S. choice beef compared with taste panel scores (eight-point scale) for tenderness (Hansen, 1972).*

Table 4.1 Simple Correlations of the Armour Tenderometer with Taste Panel Tenderness and Warner-Bratzler Shear Force.

Authors and Types of Samples	Taste Panel	Warner-Bratzler
Hansen, 1972		
U.S. Choice	−0.77	0.42
U.S. Good	−0.69	0.30
Parrish et al., 1973		
1 day postmortem	−0.40	0.19
3 days	−0.28	0.47
7 days	−0.09	−0.12
Carpenter et al., 1972		
Overall population	−0.22	0.25
U.S. Good + Choice	−0.26	0.25
U.S. Good	−0.15	0.17
U.S. Choice	−0.35	0.35
Huffman, 1974		
Overall population	−0.05	0.22

4.1). The Armour Tenderometer offered an improvement on USDA grading for the prediction of meat tenderness (Carpenter et al., 1972; Huffman, 1974), but still this was not sufficiently reliable to be of value (Parrish et al., 1973), and the Armour Tenderometer may be regarded in a rather derogatory manner by those who remember it; however, no serious student of the technology would make this mistake.

Rheology of Probe Penetration

The operation of needle probes was examined in detail by Morrow and Mohsenin (1976), who concluded the yield pressure (*PF*) is given by

$$PF = \frac{4W}{\pi D^2} \frac{1}{1 + \mu \cot \Theta/2}$$

where W is the load applied to the needle, D is the largest contact circle up the conical tip of the needle, μ is the coefficient of friction between the meat and the needle, and Θ is the conical angle at the tip of the needle. Thus, for any depth (d) of penetration relative to the conical tip,

$$PF = \frac{W}{\pi d^2 (\tan \Theta/2)^2} \frac{1}{1 + \mu \cot (\Theta/2)}$$

Although these relationships may hold for a relatively simple situation (such as a homogeneous dry material), there are many more complexities in meat. The coefficient of friction is likely to be affected by the free fluid within the system, which is a function of pH (Currie and Wolfe, 1980), but pH also may have direct effects on meat tenderness. In a practical situation, the interactions of pH, penetration force, and meat tenderness are difficult to predict. Thus, the coefficient of friction sometimes might enhance the predictive value of the Armour Tenderometer, while, at other times, it might detract.

One of the essential design features of the Armour Tenderometer is that it reduces the penetrometer operation to a single value, the maximum force required to overcome the mechanical integrity of the meat. This sacrifices a considerable amount of information on the force-deformation relationship and the time to reach peak force. This treats meat as a homogeneous system whose strongest elements are likely to be encountered during initial penetration. But our knowledge of muscle structure may suggest otherwise: a transect through the meat is a stereological sample of the tissue within, and both muscle and connective tissue have a nonrandom arrangement. Thus, it becomes a topic of paramount importance in invasive meat probe technology to understand how a probe moves through meat. Whether we are attempting to detect meat toughness rheologically or attempting to obtain smooth measurements of tissue reflectance, the way the probe moves is critical. Simply using a maximum value (such as peak force) in a signal is a risk, because it cannot be distinguished from a transient peak caused by noise (Voisey, 1976).

Dynamic Analysis of Probe Movement

Using the terminology for fat-depth probe mechanics introduced in Chapter 1, data collected at a constant rate (68 Hz) with an optical probe may be used to explore some of the mechanical properties of meat. The depth vector may be reassembled as a histogram, counting the number of data in each millimeter of depth, so that positions at which the probe pauses as it encounters strong resistance are identified by the number of data that accumulate without any advance in depth. The actual numbers are unimportant (because they are determined by the data collection rate), and they may be normalized (adjusted so that the height of the tallest histogram cell = 1). This enables us to compare in the same graphics frame the microstructural resistance at a particular depth in the meat with what the optical window of the probe encounters at the same depth. For example, if the probe tip encounters resistance from a strong seam of connective tissue at a certain depth in the meat, we will expect to see the data accumulate in the histogram cell for that depth (as the probe briefly stops moving forward) close to where the window detects strong connective tissue fluorescence (using apparatus similar to that shown in Figure 3.12).

In Figure 4.15, the way to read the results of this graphical analysis is as follows:

(*1*) The normalized optical signal for the UV fluorescence of connective tissue is shown by solid squares. Each square shows one measurement, but, often, they are superimposed.

(*2*) The histogram columns across the bottom of the graphics frame show the normalized frequency of data collection in each millimeter of meat depth. The higher the histogram column, the longer the probe pauses at the depth indicated for that column.

(*3*) At the top of the graphics frame, above the normalized data, are rectangular blocks whose *x*-axis coordinates show the depth at which the probe paused. Most blocks are narrow (where the probe stops at

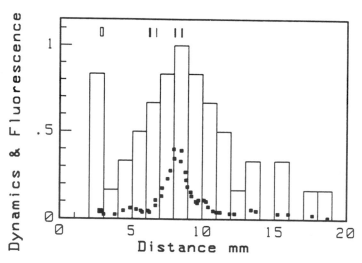

Figure 4.15 *Optical signal for the UV fluorescence of connective tissue (solid black squares) compared with the dynamics of a way-in transect into beef semitendinosus. Histogram columns show the accumulation of data at 68 Hz at each millimeter of depth, while the rectangular boxes above the normalized graphics show pauses and reversals in probe movement.*

one depth), but a wide block indicates a long sequence of disordered data in the reverse direction to the overall direction of the probe.

These features may be compared to obtain a detailed dynamic analysis of the probe vector, as in the following examples:

(*1*) A tall histogram column without a matching block at the top of the frame shows where the probe is moving slowly but always follows the overall direction of movement.

(*2*) A tall column topped by a narrow block shows where the probe stops momentarily.

(*3*) A column topped by a wide block shows where the probe bounces at it and encounters a strong elastic structure.

Features such as these may be related visually to the optical signal (solid squares) describing connective tissue distribution at the appropriate depth in question.

Not all meat samples show logical relationships between their internal connective tissue structure and probe penetration dynamics, but enough do (20 to 30%) to make it credible. Figure 4.15 shows a number of commonly occurring features in the first part of a transect. Histogram columns for the first two millimeters are empty to make sure of catching the signal on the point of entry into the meat. In other words, the zero position for depth is 2 mm above the surface of the meat. In the first column, between 2 and 3 mm in depth, the initial resistance of the meat causes a delay in probe movement (the histogram column is high, thus indicating numerous cycles of data collection without any advance into the meat), and a bounce or reversal of the probe (the *x*-axis width of the box above the column shows the limits of the disordered sequence). Being at a constant position in the meat while the probe is halted, the optical signal is constant, and most of the solid squares showing the signal are superimposed on each other.

In Figure 4.15, the solid squares show that this particular beef sample has a seam of connective tissue, peaking at 8 mm depth. The histogram columns indicating the cycles of data collection per millimeter show a matching profile as probe velocity decreases during penetration of the seam, then recovers afterwards. Above the histogram columns, the relatively narrow boxes show four positions at which the probe halts momentarily or bounces during penetration of the seam.

Figure 4.16 shows the full depth of a transect into a beef semitendinosus, with a resulting compression in the *x*-axis relative to Figure

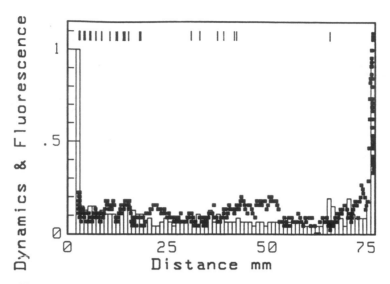

Figure 4.16 *A complete way-in transect into beef semitendinosus muscle.*

4.15. The pause at initial penetration is similar to that in Figure 4.15, but, in this example, the major seam of connective tissue is much deeper, at 75 mm. The rise in the optical signal at 75 mm is accompanied by a decrease in probe velocity and numerous short pauses in probe velocity. But no relationships between connective tissue distribution and probe dynamics are evident in the middle part of the transect (between 4 and 74 mm). The fluorescence signal shows only one potential source of meat toughness—from connective tissue—and there are numerous other factors that are invisible here, including marbling fat, muscle fiber direction, sarcomere length, and pH-related dryness. Thus, relationships between probe dynamics and connective tissue distribution may only be detectable when connective tissue is abundant, as shown in Figure 4.17, where most peaks in connective tissue fluorescence are matched by an increase in the height of the histogram columns showing deceleration of the probe. The beef on which these measurements were made had been adequately conditioned. If this had not been the case, one would expect to find an increase in the number of sites where the histogram columns indicated toughness, but the fluorescence signal was low. Conversely, when the probe window passes close by a connective tissue seam but does not pass through it, one would expect to see a peak in the fluorescence signal without any dynamic evidence of the probe slowing down.

Pauses and reversals in probe movement predominate on the initial

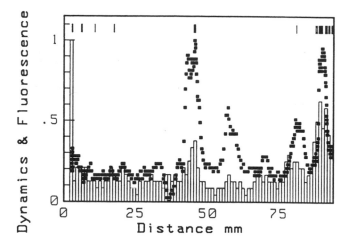

Figure 4.17 *A way-in transect of beef semitendinosus with a high connective tissue content.*

penetration of connective tissue seams. For example, Figure 4.18 shows a way-in transect where the probe encounters a very thick seam of connective tissue at 40 mm depth, but the velocity returns to normal once the initial penetration had been made. The probe may have passed through the seam at an angle, so that the seam remained in the fiber-optic window after penetration.

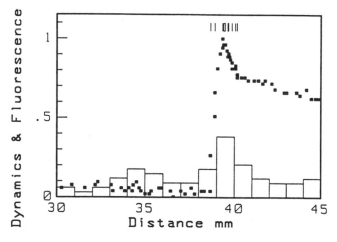

Figure 4.18 *A way-in transect from beef semitendinosus showing an increase in probe velocity after deceleration and pausing at a thick seam of connective tissue.*

Depth Vector Information

Dynamic analysis shows that irregularities of the depth vector sometimes contain useful information, as shown by the boxes at the tops of
Figures 4.15 to 4.18. When the probe tip hits resistance and the advancing probe pauses momentarily, or even bounces back a little, this
generates a disordered set of numbers in the otherwise smoothly increasing way-in depth vector.

For the calculation of parameters to describe the extent and magnitude
of disorder, the depth after a disorder may be used as a reference from
which to determine the length of the disorder. For example, consider an
imaginary sequence ($n = 6$) of integer depth measurements, 1, 2, 3, 3,
4, 5 mm. Knowing that measurements are taken at a constant rate, it is
evident that the probe pauses for one measuring cycle at 3 mm. This is
logged as a disorder length of 4 $-$ 3 = 1 mm, and an incidence of 1/6.
Another imaginary sequence ($n = 7$) where the probe bounces at 3 mm,
such as 1, 2, 3, 2, 3, 4, 5 mm is logged as length 4 $-$ 2 = 2 mm, and
an incidence of 2/7. Real numbers, the full result of analogue to digital
conversion without rounding, are used for actual calculations. Limitations to this approach may be created by rounding errors and signal noise
(signal $\pm$ mm), but, normally, they are not a problem.

Overall, the information provided by the proportion of disordered data
(number of data in reverse sequence to the direction of the vector divided
by total number in vector) conveys more information than the mean
length of the disordered sequences. In way-in penetration of beef shank
slices versus slices of psoas, the proportion of disordered data is higher
in the shank ($t = 3.17$, $P < 0.005$), whereas there is no significant
difference in the mean length of the disordered sequences (Swatland,
1991). Similarly, in beef shank versus sirloin, the proportion of disordered data is higher in the shank ($t = 3.02$, $P < 0.005$; Swatland, 1992).

In a hand-operated probe, where the prime mover is the human arm,
the penetration of the probe through the meat has a complex pattern. Our
arm movements when pushing a probe are influenced by a variety of
factors, including neuromuscular spindle and cerebellar reflexes and
proprioceptors in wrist, elbow, and shoulder. The higher centers of the
brain also are likely to exert a considerable effect. Imagine an operator
probing numerous very tough carcasses and then encountering a very
tender one: we would expect the operator to push harder for the tender
carcass if it is the last to be tested, rather than first. Thus, we may
anticipate a complex interaction between probe usage and disordered
depth vectors, as shown in Table 4.2.

Table 4.2 *Simple Correlations among Electromechanical Parameters in Using a Handheld Probe (Swatland, 1994).*

	Way-out Velocity	Way-in Disorder	Way-out Disorder	Way-in Mean Disorder Length	Way-out Mean Disorder Length
Way-in velocity	0.57[b]	−0.63[c]	−0.29	−0.03	0.19
Way-out velocity		−0.32	−0.42[a]	0.04	0.39
Way-in disorder percent			0.11	0.30	−0.14
Way-out disorder percent				0.20	−0.32
Way-in mean disorder length					0.25

[a] $P < 0.05$.
[b] $P < 0.01$.
[c] $P < 0.005$, $n = 25$.

Looking at the correlations among the electromechanical parameters in Table 4.2, it is evident that way-in and way-out velocity are likely to be correlated, while velocity and disorder may be negatively correlated for matching way-in and way-out directions. In other words, if a probe is pushed into the meat rapidly, it is likely to be withdrawn rapidly. And when a probe is pushed or pulled rapidly, the depth vector is less likely to be disordered than when the probe is moved with less force. Obviously, the data in Figure 4.2 describe only one particular experiment, but they are typical for this type of probe.

Subjective Bias in Probe Usage

This scrutiny raises some serious and, as yet, largely unanswered questions relating to the whole concept of handheld meat probes, including the fat-depth probes that are used widely in pork grading. To what extent does an operator prejudge the force required to penetrate the meat with a handheld probe? Do human judgement and arm reflexes affect probe velocity? Does probe velocity affect the operation of the probe? In other words, because of these possibilities, subtle subjective factors may intrude into otherwise objective methods of on-line evaluation of meat.

The subjective element involved in a human operator using a handheld meat probe does not invalidate the technology, but neither may it be ignored. Does meat temperature or the direction that measurements are made relative to the muscle fibers affect the way in which an operator uses a probe? Table 4.3 shows some typical effects for sample temperature ($<0°$ versus $5°C$) and direction of probe penetration relative to the longitudinal axes of muscle fibers. The pattern for way-in velocity matches that for way-out velocity, being faster across the fibers than along them and faster for meat at $<0°$ than at $5°C$. This seems reasonable if more force is required to push or pull the probe across the fibers or in stiff meat at the point of thawing, because, once the greater initial resistance is overcome, one would expect the stronger force to create a higher velocity.

The analyses of depth vectors from handheld probes reveals some potential problems, but it also offers solutions, thus helping to improve the technology. For example, depth vector analysis could be used in a training program for probe operators. As well as producing a fat-depth or carcass grade, the training probe also could provide a feedback to the user, either audible or visual, teaching the operator how to use the probe properly. Humans have a strong ability to learn from novel forms of feedback, which is exploited in fields as diverse as driver training and biofeedback, and this could be used to train and test probe operators. Alternatively, depth vector analysis might be used with existing on-line probes to give a warning that probe velocity or juddering had been beyond accepted limits and that the measurement should be repeated. The acceptable limits of performance may be found from relationships such as those shown in Figure 3.25.

Sometimes the disorder of the depth vector is correlated with the composition of the meat: for example, disorder may be correlated with biochemically determined collagen content ($r = 0.48$, $P < 0.01$; Swatland, 1993). But to exploit such relationships in the on-line evaluation of meat would require automation of the probe delivery system to remove or standardize the subconscious element of subjectivity.

MIRINZ Torque Tenderometer

A variety of different configurations have been used to test the rheology of meat in the laboratory (Voisey, 1976), and a few have been used on-line. A recent development (Figure 4.19) is a handheld tenderometer for on-line use that consists of two sets of concentric needles. The outer two rings of longer needles are fixed, while the inner ring of shorter needles is rotated by a synchronous motor, and the torque and degree of rotation to tear the meat are determined. The peak torque and the torque at $60°$ are correlated with cooked meat tenderness

Table 4.3 *Comparing Measurements at $<0°$ versus $5°C$, and Measurements Parallel to the Fibers versus Perpendicular to the Fibers (Swatland, 1994).*

	$<0°$ Parallel	$5°$ Parallel	$5°$ Perpendicular
Way-in velocity cm sec^{-1}	2.14 ± 0.56[c]	1.69 ± 0.33[c]	2.39 70.54
Way-out velocity cm sec^{-1}	2.23 ± 0.64[b]	1.88 ± 0.37[c]	2.57 ± 0.55
Way-in disorder	0.16 ± 0.11	0.16 ± 0.06	0.18 ± 0.09
Way-out disorder	0.14 ± 0.09	0.14 ± 0.06[c]	0.11 ± 0.03
Way-in mean disorder length mm	0.07 ± 0.03[a]	0.05 70.01	0.05 ± 0.03
Way-out mean disorder length mm	0.06 ± 0.01[a]	0.07 ± 0.01	0.07 ± 0.02

[a] $P < 0.025$.
[b] $P < 0.01$.
[c] $P < 0.005$ for differences between adjacent columns.

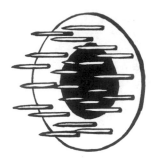

Figure 4.19 MIRINZ Tenderometer (MIRINZ, 1991).

(MIRINZ, 1991). Phillips and Loeffen (1993) found the MIRINZ Tenderometer could be used to identify cold-shortened beef (Figure 4.20). Raw meat is tested with the needles parallel with and coaxial to the muscle fibers, and the peak torque is correlated with tenderometer tests after cooking, $r = 0.90$ (Phillips and Loeffen, 1993).

Australian Tender-Tec Probe

Another new device for measuring meat toughness was introduced by Gordon (1994). The Tender-Tec developed by the Australian Meat

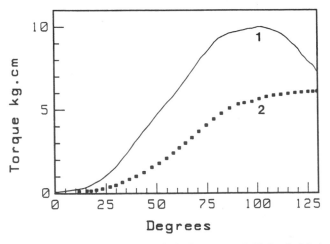

Figure 4.20 Separation of cold-shortened (1) from normal (2) beef with the MIRINZ Tenderometer used on raw meat with the needles parallel and coaxial with the muscle fibers (Phillips and Loeffen, 1993).

Research Corporation is a battery-powered, handheld penetrometer similar in concept to an optical fat-depth probe with a single needle a few millimeters in diameter. However, instead of making optical measurements as a function of depth in the carcass, a strain-gauge measures resistance to probe penetration and plots this as a function of depth. For beef carcasses with a wide range in age and, hence, connective tissue toughness, the Tender-Tec is correlated with meat toughness, $R^2 = 0.78$.

ULTRASONICS

Basic Properties

Ultrasonics achieved sudden prominence as a method for on-line meat evaluation when the U.S. meat industry announced plans for an extensive program of research and development in this technology (Kester, 1991). During instrument grading, a mechanism will be attached to the carcass that will enable an ultrasonic estimation of lean yield, marbling, and skeletal maturity (Novakofski, 1993); however, this is a long-term initiative that will not be completed until the end of the century (Cross and Savell, 1994).

Ultrasound is sonic energy at frequencies beyond the audible range (>20 kHz). Sound moves by compression waves that, when they pass from one medium to another, may be reflected or refracted. The amount of energy reflected depends on the difference in density between the two media and on the angle of incidence at which the beam of ultrasound encounters the interface between the media. If the density difference between the media is large, then reflection is high. If the angle of incidence is near $90°$, then reflection is high. Thus, a beam of ultrasound perpendicular to a bone or an air space in the meat is nearly all reflected.

Acoustic impedance is the ratio of ultrasonic pressure to the resulting flux, but, since it is a complex ratio containing both resistive and reactive components, a simpler ratio is more useful. The characteristic impedance is the product of the density of a medium and the velocity of ultrasound through it. Energy is absorbed as ultrasound moves through a medium so that the flux is attenuated; thus, the attenuation constant is the relative rate of decrease in ultrasound amplitude in the direction of movement. The relevance to on-line meat evaluation is that attenuation increases as a power of frequency. Thus, a high frequency has a poor penetration; however, low frequencies have poor resolution of particles within the meat. High frequencies are more liable to be scattered than low frequencies, but this loss of resolution is balanced by the fact that high frequencies are more easily focused than low frequencies. In summary, the choice of frequency (usually from 1 to 15 MHz) is a compromise between penetration and resolution, so that it is difficult or impossible to resolve small particles at great depth in the meat.

A-mode and B-mode

Ultrasonic imaging involves pulses of ultrasound generated from a piezoelectric crystal, with reception of the echoes by the same or another crystal. The simplest form of imaging is an A-mode (amplitude mode) display on an oscilloscope with a stationary transducer. The oscilloscope sweep is triggered when the pulse is generated so that the length of time for echoes to return may be read as the equivalent distance on the x-axis of the oscilloscope.

The A-mode may be used to measure the fat content of meat. The velocity of ultrasound in meat decreases when marbling levels are high, at a rate of 2.69 m/(s-% fat). Using a nonlinear regression model, intramuscular fat content of beef may be predicted with 90% accuracy for fat levels over 8% and with 76% accuracy for under 8% fat (Park et al., 1994a). However, at present, the technique requires cubes of meat to be removed from the carcass and accurately located between two aluminum plates, and the direction of the muscle fibers also must be controlled. It may be possible to replace the lower aluminium plate reflector by a natural reflector in the live animal (such as a boundary between fat and lean at a known depth), but whether the method can be made to work without accurate control of the depth of tissue remains to be seen. Another approach is a Fourier analysis of the frequency spectrum of the A-mode echo, which allows predictions at fat levels below 4% (Park et al., 1994b).

Instead of the amplitude of the echo being used to drive the y-axis of the oscilloscope, as in the A-mode, the amplitude may be used to make the oscilloscope spot more intense or bright. Thus, if the spot intensity is reduced to the threshold level expected for an echo, when an echo appears, it is seen as a bright spot on the screen, rather than as a peak in the y-axis direction. The B-mode (brightness mode) display commonly used to obtain images of fat depth and rib-eye areas from meat animals uses this same principle, but the transducer is moved in steps synchronized with steps in the y-axis of the display device. Thus, the bright spots indicating reflective structures in the carcass build up a two-dimensional image.

B-mode images of beef muscle may contain a speckle pattern related to small inclusions in the muscle, mostly marbling. Autocorrelation analysis of the speckle pattern may provide predictions of marbling (Liu et al., 1993), but results are somewhat variable (from $R = 0.82$ to 0.21). Autocorrelation is used to detect cyclic activity in a signal. A copy of the signal is delayed, then multiplied by the undelayed signal. After smoothing with a low-pass filter, meaningful signal peaks from marbling persists, while noise is lost. Another approach is to use video pattern recognition coupled with a neural network (Brethour, 1994).

Danish AUTOFOM for On-line Pork Grading

Instead of moving the transducer in step with the display device to generate an image, as in B-mode ultrasonics, it is also possible to move the whole carcass past an array of transducers mounted in a U shape to match the profile of the dorsal region of the carcass. This creates a three-dimensional image, as in the Danish AUTOFOM produced by SFK (Mannion, 1991). The AUTOFOM may process 900 carcasses per hour and is located immediately after the scalding tank when carcasses have maximum flexibility for being pulled through the transducer U frame. As well as predicting lean yield, the AUTOFOM also provides information on the distribution of meat within the carcass, which, when coupled with individual carcass identification encoded into the gambrel, allows carcasses to be sorted for maximum utilization in meat cutting. Other ultrasonic fat-depth systems are also available, such as the CSB Ultra-Meater (Lay, 1992).

ROBOTIC PROBES

Handheld probes are a prime candidate for automation as soon as labor costs, reliability, and the potential for human repetitive strain injury are taken into account. The Danish Carcass Classification Center (Plates 2 and 3) is a good example of a dedicated machine that outperforms a handheld system and effectively deals with the problem of variation in carcass size and shape (Chapter 1). It is difficult to find a single anatomical position that gives a representative average of the sub-cutaneous fat depth or meat quality for the whole carcass, and the Danish solution is to make a series of measurements with a battery of probes. If one probe malfunctions, then the others may compensate, using a neural network (Thodberg, 1992). An alternative approach is to use a single probe delivered by a robot that may be programmed to undertake a variety of repetitive measurements. In other words, a robotic system offers far greater flexibility than a dedicated machine. The relevance to the electromechanical properties of meat is that the robot must interact with the rheological and mechanical properties of the carcass. For example, penetration of the probe into the meat contains information similar to that obtained by dynamic analysis.

Goldenberg and Ananthanarayanan (1993) developed a robotic probe to be used in conjunction with a carcass clamping subsystem. The probing subsystem is composed of an ultrasonic scanner to determine the locations of the carcass skeleton, a mid-split scanning subsystem to find the edge of the split vertebral column, and a pneumatic probe insertion and retraction subsystem. For pork, the robot faces the back of the recently split pork carcass as it progresses along the line and scans downward from lumbar to thoracic regions. The statutory point for probe grading in Canada, upon which regressions are based to calculate meat yield, is 7 cm from the midline between the third and fourth last ribs. The mid-split scanner has a reflecting-type proximity sensor composed of an LED and a photodetector, with sensor repeatability of approximately 5 μm at a proximity range of 25 mm. The ribs are counted ultrasonically at a frequency that optimizes sound transmittance and echo location, counting in a posterior to anterior sequence as the scanner moves down the carcass. Water jets provide continuous ultrasonic coupling to the carcass. The quality of the ultrasonic echo is very sensitive to orthogonality between the source plane and the carcass surface, so the sensor is mounted at the apex of a tripod, while the base follows the surface curvature of the carcass, gliding smoothly on the water jets.

The potential advantage of a robot is that it is inherently flexible, lending itself to multiple applications such as measuring different parts of the carcass, different species, and variable numbers of measurements to attain a specified confidence level. The rib-counting ultrasonic sensor already provides information on fat depth and improves the sampling base by taking a long scan down the carcass, thus enabling a meat-quality probe to be used at multiple measuring sites.

Replacing human knowledge and common sense with computer software requires a new approach to carcass anatomy. Although we may recognize the visual features of a rib cage without really knowing how we do so, the robot needs to know the shapes, sizes, and angles of the whole anatomical system to navigate by itself ultrasonically. Even in a region of the carcass so thoroughly known as the thoraco-lumbar junction, every feature must be described and measured in a new way. In pigs, there is considerable variation in the distribution of thoracic vertebrae (numbering from 13 to 17) and lumbar vertebrae (from 5 to 7) caused by breed variation in animal body length. Whether or not paraxial ossification centers develop into ribs articulating with a thoracic vertebra, or into transverse processes fused to the centra of the lumbar vertebra, depends on morphogenetic gradients in the embryo. Thus, the common procedure to locate a handheld probe is to use the last rib as a reference point, without counting all the ribs. Thus, the first lumbar vertebra with a transverse process may be identified correctly, but the number designation of the last thoracic vertebra remains unknown, because of the potential variation in the number of thoracic vertebrae.

The robot software must take into account the anatomical variability among pork carcasses in order to locate the designated probing site. This may be done using an intelligent algorithm based on a methodology that mimics human commonsense reasoning. A technique known as qualitative reasoning may be used to analyze imprecise or incompletely specified data to generate control decisions for the motion of the robot; however, this requires estimates of the possible values of the navigational data (rib thickness, rib angle, intercostal distance, rib curvature, etc.). Based on these approximate values, the intelligent control system then may be able to reach the designated probing location reliably and accurately.

Rib curvature may be navigated using two ultrasonic sensors in parallel and by multiplexing the analogue input from the ribs to an analogue to digital convertor. A total of eight different points may be recorded on a rib within 1.5 seconds by the ultrasonic scanner moved by the robot end-effector. This is made possible by high-speed, real-time parallel computational architectures. For rib navigation, the robot may move at an average velocity of 0.5 m sec^{-1}, using a twelve-bit analogue to digital convertor with a conversion time of 5 μs and a parallel RISC coprocessor capable of 5 MFLOPS. Using the eight data points, the model of the rib curvature is reconstructed (a seventh-order approximation) on-line by the robot controller to identify the designated probing site.

REFERENCES

Bate Smith, E. C. 1939. *J. Physiol.*, 96:176.

Bendall, J. R. 1973. Postmortem changes in muscle, in *Structure and Function of Muscle, Vol. 2*, part 2, G. H. Bourne, ed., Academic Press, New York, pp. 244–309.

Blackmore, D. K. and J. C. Newhook. 1982. *Meat Sci.*, 7:19.

Bouton, P. E., A. L. Ford, P. V. Harris and F. D. Shaw. 1980. *Meat Sci.* 4:145.

Brethour, J. R. 1994. *J. Anim. Sci.*, 72:1425.

Carpenter, Z. L., G. C. Smith and O. D. Butler. 1972. *J. Food Sci.*, 37:126.

Cross, R. and J. W. Savell. 1994. *Meat Sci.*, 36:19.

Currie, R. W. and F. H. Wolfe. 1980. *Meat Sci.*, 4:123.

Davis, C. E., W. E. Townsend, and H. C. McCampbell. 1978. *J. Anim. Sci.*, 46:376.

Gilbert, K. V. and C. E. Devine. 1982. *Meat Sci.*, 7:197.

Goldenberg, A. A. and S. Ananthanarayanan. 1993. *Food Res. Internat.*, 27:191.

Gordon, T. 1994. The Tender-Tec approach to determine important carcass attributes, in *National Beef Instrument Assessment Plan, May 25–26, 1994*, Chicago, IL: National Live Stock and Meat Board.

Hansen, L. J. 1972. *J. Texture Stud.*, 3:146.

Huffman, D. L. 1974. *J. Anim. Sci.*, 38:287.

Kester, W. 1991. *Beef* (Minneapolis) 28(4):14.

Lay, K. 1992. *Fleisch*, 46:447.

Liu, Y., D. J. Aneshansley, and J. R. Stouffer. 1993. *Trans. Am. Soc. Engin.*, 36:971.

Loeffen, M. 1992. Clip-on stimulation monitor launched, in *Added Value MIRINZ Newsletter*, 8, Hamilton, NZ: Meat Industry Research Institute of New Zealand.

Mannion, P. 1991. *Meat Internat.*, 3(3):21.

Marsh, B. B. 1954. *J. Sci. Food Agric.*, 5:70.

MIRINZ, 1991. *MIRINZ Meat Research, Annual Report 1991–1992*. Meat Industry Research Institute of New Zealand, Hamilton, NZ, pp. 20–21.

Morrow, C. T. and N. N. Mohsenin. 1976. *J. Texture Stud.*, 7:115.

Newhook, J. C. and D. K. Blackmore. 1982a. *Meat Sci.*, 6:221.

Newhook, J. C. and D. K. Blackmore. 1982b. *Meat Sci.*, 6:295.

Novakofski, J. 1993. Instrument grading – Development of an ultrasound-based system, in *Dimensions of Research from Strategies to Users*, Chicago, IL: National Live Stock and Meat Board, p. 18.

Park, B., A. D. Whittaker, R. K. Miller, and D. S. Hale. 1994a. *J. Anim. Sci.*, 72:109.

Park, B., A. D. Whittaker, R. K. Miller, and D. E. Bray. 1994b. *J. Anim. Sci.* 72:117.

Parrish, F. C., D. G. Olson, B. E. Miner, R. B. Young, and R. L. Snell. 1973. *J. Food Sci.*, 38:1214.

Phillips, D. M., and M. P. F. Loeffen. 1993. *Meat Focus Internat.*, 2:16.

Savell, J. W., G. C. Smith, and Z. L. Carpenter. 1978. *J. Anim. Sci.*, 46:1221.

Swatland, H. J. 1975. *Can. Inst. Food Sci. Technol. J.*, 8:122.

Swatland, H. J. 1976a. *J. Anim. Sci.*, 42:838.

Swatland, H. J. 1976b. *J. Anim. Sci.*, 43:577.

Swatland, H. J. 1983. *Can. Inst. Food Sci. Technol. J.*, 16:35.

Swatland, H. J. 1985. *J. Anim. Sci.*, 61:882.

Swatland, H. J. 1986. *Can. Inst. Food Sci. Technol. J.*, 19:167.

Swatland, H. J. 1991. *Comput. Electron. Agric.*, 6:225.

Swatland, H. J. 1992. *Comput. Electron. Agric.*, 7:285.

Swatland, H. J. 1993. *Comput. Electron. Agric.*, 9:255.

Swatland, H. J. 1994. *Food Res. Internat.*, 27:433.

Sybesma, W. 1966. *Fleischwirts.*, 6:637.

Thodberg, H. H. 1992. The neural information processing system used for pig carcase grading in Danish slaughterhouses, in *Symposium on Image Analyses in Animal Science*, Research Centre Foulum, Tjele, Denmark.

Voisey, P. W. 1976. *J. Texture Stud.*, 7:11.

van der Wal, P. G. 1978. *Meat Sci.*, 2:19.

CHAPTER 5

Optical Properties of Meat

INTRODUCTION

THERE IS A wealth of information in our libraries, concerning the optical properties of skeletal muscle, but mostly at cellular and subcellular levels in relation to the mechanism of muscle contraction. For the on-line evaluation of meat, the muscle tissue is in a bulk state: millions of muscle fibers set in rigor mortis, losing their fluids and membranes. Thus, although basic scientific knowledge is essential in understanding the bulk optical properties of meat, we need a further level of applied scientific work to understand how meat probes may be used and improved for the on-line evaluation of meat.

MICROSTRUCTURE OF MEAT

The bulk optical properties of meat originate from the combined properties of large numbers of muscle fibers, and some introduction is needed for those who are unfamiliar with the histology of muscle. In live animals, a sliding interaction of microscopic filaments enables a muscle to contract, while, in meat, an ordered arrangement of fasciculi (bundles of fibers), fibers, fibrils, and filaments creates a characteristic texture that is difficult to imitate with processed plant proteins. These components of meat (fasciculi, fibers, fibrils, and filaments) constitute a descending series with respect to size. Fasciculi are the largest units, visible to the naked eye as the grain of the meat, while filaments are the smallest. The prefix myo- may be used to create the terms myofiber, myofibril, and myofilament, which are identical to muscle fiber, muscle fibril, and muscle filament, respectively.

Fasciculi

Complexity of muscle structure is related to muscle function in the live animal. It previously was thought that muscles that contracted to a relatively small fraction of their resting length had their fibers parallel to the long axis of the muscle, whereas muscles in which the strength of contraction was more important than the distance over which contraction occurred had fibers with an angular arrangement. Thus, it was thought that muscles might gain strength by leverage, reducing the contraction distance. This was a widespread idea but with little or no evidence to support it. A pennate muscle structure (a muscle with V-shaped subunits) may, in fact, serve to increase the length over which a muscle contracts.

The fascicular structure of meat is maintained by connective tissue. Thin, frequently branching collagen fibers (Plate 17) weave a tube of endomysium around each muscle fiber (Plate 18). Endomysial tubes are bound together in fasciculi by thick collagen fibers of the perimysium, similar to the septum seen in Plate 19. Finally, on the muscle surface, the outermost sheets of perimysium blend into a very thick connective tissue layer of epimysium. Sometimes the epimysium is thickened even further to form a flat tendon or aponeurosis, extending beyond the muscle as a flat sheet of connective tissue or fascia.

Fibers

If a small amount of meat is placed under a dissecting microscope and teased apart with needles, the smallest fasciculi visible without magnification are composed of bundles of muscle fibers. Muscle fibers are the basic cellular units of living muscle and meat. They are unusual cells because they are multinucleate and extremely long (commonly several centimeters) relative to their microscopic diameter (usually less than 100 μm). The muscle fibers found in most commercial cuts of meat seldom run the complete length of the muscle in which they are located. Individual fibers within a fasciculus may terminate at a point anywhere along the length of the fasciculus, tapering to an ending that is anchored in the connective tissue on the surface of an adjacent muscle fiber. Tapered endings transmit their force of contraction to the endomysium.

Apart from tapered intrafascicular endings, the diameter of a muscle fiber is approximately constant along its length. Fiber diameters increase as an animal grows to market weight, but they also increase temporarily when a fiber contracts. Thus, when measuring fiber diameters in meat, special care must be taken to avoid or correct for differences in the degree of muscle contraction. If a muscle fiber becomes detached within the live animal, because of muscle exertion or some form of muscle degeneration, or if a muscle sample is taken directly from a recently slaughtered animal, the detached fiber may be unable to relax back to its normal diameter. Detached fibers are often known as giant fibers. An example is indicated by the arrow in Plate 19, although they may become much larger than this.

Myofibrils and Transverse Striations

If muscle fibers are macerated with water in a blender, after a few seconds, the connective tissue holding the muscle fibers together is disrupted, leaving a pale red suspension of broken muscle fibers. The red color originates from myoglobin, a soluble red pigment found inside muscle fibers. In pork, most of the fibers with a high myoglobin content are grouped together in the center of their fasciculus, as shown in Plate 10. This is not observed in other species, where the distribution of fibers with a high myoglobin content is more random, as in Plate 11, where more subtle differences in myoglobin content between fibers were revealed histochemically. There is very little hemoglobin from red blood cells in carcasses of meat animals that have been properly exsanguinated.

The main features of muscle fiber structure may be seen with a light microscope if a drop of macerated muscle suspension is mounted on a microscope slide beneath a cover slip. Muscle fibers basically have a cylindrical shape, and, along their lengths there are many transverse striations. The striations become visible if the aperture of the substage condenser is reduced (the gain in contrast being offset by a loss of resolution) or, even more clearly, if a polarizing microscope is used, as in Plate 8.

With a high magnification microscope objective, myofibrils may be seen protruding from the broken ends of fibers or floating freely in the mounting medium having escaped from a broken fiber. Beneath the surface membranes of the lengths of broken muscle fibers are flattened bubble-like inclusions, the nuclei. Their DNA may be stained with dyes such as hematoxylin.

On the surfaces of fibers retaining their endomysial tubes may be seen branching capillaries, once part of the vascular bed of the muscle. Red blood cells may or may not remain in the capillaries, depending on the efficiency with which the animal was bled. If they do remain in the meat, they appear pale yellow in unstained preparations and are often distorted or piled together like a stack of coins.

When meat animals are slaughtered, normally, they are shackled and suspended from their hind limbs. Some muscles, such as the filet or psoas muscles ventral to the vertebral column, become stretched. Other muscles, such as those in the posterior part of the hind-limb, are free from skeletal restraint and may contract weakly as rigor mortis develops. If samples from stretched and contracted muscles are compared, transverse striations appear relatively far apart in the stretched muscle and close together in the contracted muscle. The distance between the transverse striations is the sarcomere length. Meat with short sarcomeres generally is much tougher than meat with long sarcomeres, provided that the connective tissue content of the samples is similar.

If a drop of saturated salt solution (NaCl) is mixed with a drop of macerated muscle suspension, the muscle fiber fragments undergo some marked changes. The fiber fragments swell and may disappear, or they may expand so violently that their interiors are extruded from their broken ends. The solubility of meat proteins in salt solutions is basic to many meat processing operations, enabling fragments of comminuted meat to be bound together to produce a processed meat product.

After prolonged maceration in a blender, muscle fiber fragments disintegrate, and their myofibrils are released into suspension. With a light microscope at high magnification, it is possible to measure sarcomere lengths, which may be seen more accurately on myofibrils than on fibers. With a polarizing microscope, the substage polarizer provides plane polarized light oriented at 0° for the illumination of the specimen, and the analyzer in the microscope tube is set at 90° to create a dark field (Hartshorne and Stuart, 1970). Under these conditions, alternate transverse striations along fibers and myofibrils oriented at 45° appear bright, thus revealing their intrinsic birefringence. Thus, having two refractive indices (fast and slow axes) allows the bright bands to rotate the polarized light strongly enough for the light to get through the analyzer. Rotation of the microscope stage allows this property to be observed in any particular fiber or myofibril as it is brought to 45°. Striations that appear bright in polarized light are termed anisotropic or A bands, while those that are dark or relatively dim are termed isotropic or I bands (Plate 8).

The transverse striations on muscle fibers originate from the precise alignment of A and I bands on myofibrils within the fiber. In most stained preparations for light and electron microscopes, A bands appear darker than I bands (the reverse appearance to that seen with polarized light). This is because the birefringent A bands contain a greater density of protein than the I bands. A bands also appear darker than I bands when unstained preparations are observed by phase contrast microscopy.

Myofilaments

At the ultrastructural level, the transverse striations of myofibrils are caused by the regular longitudinal arrangement of sets of thick filaments (10 to 12 nm in diameter) and thin filaments (5 to 7 nm in diameter). In a transverse section through overlapping thick and thin filaments of the sarcomere, each thick filament is surrounded by six thin filaments, although this hexagonal lattice may change to a tetragonal lattice when sarcomeres are stretched. When a muscle fiber contracts, the thick filaments slide between the thin filaments so that the I band gets shorter. The length of the A band remains constant. This is the sliding filament theory of muscle contraction, proposed in the early 1950s by two unrelated scientists working independently but, confusingly, both called Huxley (Andrew Huxley in Cambridge, England, and Hugh Huxley in Cambridge, Massachusetts). If a muscle is at its resting length, the gap between opposing thin filaments at the midlength of the sarcomere causes a pale H zone in the A band.

Other features of the myofibril are detectable by light microscopy under optimum conditions, but these details are seen more clearly by electron microscopy. A thin Z line or Z disk occurs at the middle of the I band. The sarcomere length is generally taken as the distance from one Z line to the next. The Z line is a complex structure formed from woven protein filaments, and it extends as a partition across the myofibril, anchoring the thin filaments.

In electron micrographs of meat, we seldom see the fine details evident in muscle samples treated to preserve the details of the living system. Plate 9 shows a typical longitudinal section of pork, showing the A and I bands and Z lines of myofibrils. The alignment of A and I bands in adjacent myofibrils has been disrupted. Individual thick and thin filaments are visible in the transverse section of a disintegrating pork myofibril in Plate 16, but not much remains of the hexagonal pattern.

Energy for muscle contraction originates by the hydrolysis of phosphate from ATP, although the transduction from chemical to mechanical energy may be delayed until the resulting adenosine diphosphate (ADP) and inorganic phosphate are released by myosin when it recombines with actin. Contraction by filament sliding may be achieved by a rowing action of numerous cross bridges that protrude from the thick filaments. The cross bridges are formed from the heads of myosin molecules whose backbones are bound into the thick filament. However, the conformational change that causes the cross bridge movement does not seem to be a simple angular movement of the cross bridge, as was originally supposed, and the movement might originate elsewhere in the molecule, whose structure was first established three-dimensionally by Rayment et al. (1993).

Filament sliding and muscle contraction originate from the conformational changes of very large numbers of myosin molecules. Each individual stroke by a myosin molecule head takes about one millisecond and produces a 12-nm movement. Although this is a very small distance, many thousands of sarcomeres are arranged in series, and, in a very short time, the sum of all these small distances may produce a movement of many centimeters. The myosin head only releases its grip on the actin, before swinging back for another power stroke against another actin molecule, if it is recharged with another ATP molecule. Thus, when muscles are converted to meat and no more ATP is available, thick and thin filaments lock together wherever they overlap. This prevents any further filament sliding, and the muscle becomes almost inextensible, set in rigor mortis.

Actin molecules are arranged in an elongated double helix in the thin filament. The two grooves of the helix are important because they contain other proteins that control muscle contraction. Troponin responds to the presence of calcium ions and causes tropomyosin to change its depth in the groove. This allows myosin heads to reach the actin molecules against which they can move to cause contraction. Thus, the release of calcium ions triggers muscle contraction. In a relaxed muscle, the calcium ions are resequestered by the sarcoplasmic reticulum, which is a complex series of membranous structures around the myofibrils.

CAUSES OF pH-RELATED PALENESS

At least one-quarter of the pork retailed in the United States has been judged subjectively as PSE (Kauffman et al., 1993). Even though on-line pH_1 values (as shown in Table 2.2) are difficult to find, there is no doubt that PSE is of major importance in the United States. Since there is overwhelming evidence that PSE is a pH-related effect, one might expect to find a large body of scientific information explaining how low pH causes meat paleness and how high pH causes darkness. Yet this is not the case. We have a few published theories, plus some recent experiments, but nothing that yet approaches a reliable scientific consensus. Thus, a critical approach to what follows is required, since the conclusions may change as new ideas and evidence come to light.

An early attempt to explain the paleness of low-pH meat was presented by Bendall and Wismer-Pedersen (1962) who proposed that light scattering at a low pH is caused by protein denaturation. Light scattering is a complex phenomenon, covered in detail in many physics texts. At the level that concerns us, however, one might compare the denaturation of meat proteins by acid to the denaturation of egg albumen proteins by heat (Bendall, 1962). Thus, as an egg is cooked, the albumen changes from a transparent liquid to a soft, white solid. By analogy, as the acidity of meat develops after slaughter, some of the transparent, soluble meat proteins may be denatured, and the resulting increase in light scattering may increase the paleness of the meat. Evidence in support of this hypothesis is quite convincing because the precipitated protein is detectable histologically in severely PSE pork (Lawrie et al., 1958; Bendall and Wismer-Pedersen, 1962). When stained with hematoxylin, the protein precipitate forms irregular dark bands, mostly perpendicular to the longitudinal axis of the muscle fiber, like sarcomeres, but much wider ($> 20\,\mu$m). Sometimes the irregular bands of precipitate cause the plasma membrane of the muscle fiber to bulge outwards. The myofibrils of affected muscle fibers generally appear to be quite normal, so Bendall and Wismer-Pedersen (1962) concluded that the precipitate of denatured protein is derived from soluble sarcoplasmic proteins. If correct, such an effect certainly would increase the paleness of meat by causing an increase in light scattering.

From a critical perspective of the Bendall and Wismer-Pedersen (1962) hypothesis, however, the main concern is with artifactual precipitation: a protein may be denatured without always being precipitated to scatter more light. There is no doubt that a precipitate of some sort was observed in muscle fibers from PSE pork, but the fibers had been fixed in formaldehyde, dehydrated (probably with ethanol), and embedded in celloidin (normally ether is used to pretreat the samples before adding the cellulose nitrate). A demonstration of precipitation bands in untreated muscle fibers from PSE meat (using phase or interference contrast microscopy) would alleviate this concern. Until such time, it must be considered at least possible that, although denatured by acid in the meat and highly susceptible to precipitation by formaldehyde and ethanol during histological preparation, the sarcoplasmic proteins may not actually have been precipitated into light-scattering bands across the fibers in the original PSE meat. Despite these concerns, the hypothesis probably is correct, and precipitation of denatured sarcoplasmic proteins by a low pH probably contributes to the extreme paleness of severe PSE. But how does protein precipitation explain the differences in light scattering between normal and DFD pork, neither of which have histologically detectable bands of precipitated proteins?

Another early hypothesis suggested by Hamm (1960), and supported by Offer and Trinick (1983), proposed that shrinkage of myofibrils at a low pH increases the refractive index difference between the myofibrils and their surrounding sarcoplasm. This would act to increase the scattering of light from the myofibrillar surface, and evidence of this was given in a micrograph obtained by scanning confocal light microscopy and published in a review by Offer et al. (1989). No evidence of light scattering from molecular features within the myofibril was detected by Offer et al. (1989). This second hypothesis (pH-related changes in refraction and/or reflection of light at the myofibrillar surface) is attractive because it may explain the differences in light scattering between normal and DFD meat.

Although Offer et al. (1989) dismissed the internal structure of the myofibril as a source of pH-related light scattering, this may be somewhat premature. Myofibrils account for much volume in pork, starting at approximately 80 % and declining to 50 % as the pH declines postmortem (Bendall and Swatland, 1988), so that even small optical changes could have a major impact. Increased myofibrillar refractive index caused by low pH might increase scattering by increasing the refractive angular deflection of light passing through myofibrils. Following Bendall's (1962) analogy of muscle protein denaturation with heat-induced denaturation of egg white, a salient point is that the refractive index of egg albumen increases when it is cooked (Bolin et al., 1989). Refractive index is difficult to investigate experimentally but, fortunately, myofibrils are strongly birefringent, as indicated by the naming of A (anisotropic) and I (isotropic) bands, and this enables them to be investigated with polarized light (Engelmann, 1878). Measurements on individual muscle fibers show that the optical path difference (between rays following different refractive pathways allowed by birefringence) tends to increase when pH is decreased (Swatland, 1989a, 1990a). Thus, although a complete explanation of light scattering in meat probably involves the first two hypotheses outlined above (Bendall and Wismer-Pedersen's hypothesis on protein precipitation and Hamm's hypothesis on scattering at the myofibrillar surface), the only hypothesis with readily available experimental data is the third, involving myofibrillar birefringence.

Protein Precipitation Transmission Value

The protein denaturation hypothesis of Bendall and Wismer-Pedersen (1962) is similar in concept to a laboratory method of PSE evaluation proposed by Hart (1962, with an English translation by Dekker and Hulshof, 1971). Soluble proteins are extracted by homogenization from a 5-gm sample of meat in 15 ml water. After centrifugation and filtering, 1 ml of filtrate is transferred into a cuvette containing 5 ml citrate phosphate buffer at pH 4.58 and kept in a water bath at 20° for thirty minutes. Transmittance then is measured at 600 nm and compared to a blank containing 1 ml meat filtrate plus 5 ml water.

Meat with little or no intrinsic protein denaturation produces a filtrate with a high content of soluble protein, which subsequently is precipitated at pH 4.58 to give a low transmission value. PSE meat, on the other hand, already has undergone intrinsic protein precipitation, so that the filtrate has a low content of soluble protein, little precipitation upon acidification, and a high transmittance value. Thus, transmission values are negatively correlated with pH: for example, $r = -0.68$ for the correlation of transmission value with pH_{40} in the pork longissimus dorsi (Schmidt et al., 1971). Correlations of transmission value with pH may vary with the time that the pH measurement is made: for example, at forty minutes, twenty-four hours, and forty-eight hours after slaughter, respectively, Martin and Fredeen (1974) found $r = -0.43, -0.53$, and -0.72.

Transmission values might enable us to evaluate the extent that denaturation of sarcoplasmic proteins causes meat paleness, by examining the correlations of transmittance values with subjective color scores of pork. Extraction of meat proteins by water, as in the Hart (1962) method, may cause the extraction of some myofibrillar proteins in addition to those of the sarcoplasm (Stanley et al., 1994), but this may not invalidate the concept.

The problem originates from the population of pork carcasses that is examined. As we have already surmised, the Bendall and Wismer-

Pedersen hypothesis is applicable to a comparison of PSE with normal pork but not to a comparison of normal with DFD pork. Thus, transmission values do not distinguish reliably between normal and DFD pork (Kauffman et al., 1986), and a population containing numerous DFD carcasses will cause us to underestimate the contribution of protein denaturation to PSE.

The strongest evidence in support of the hypothesis of Bendall and Wismer-Pedersen comes from studies by Eikelenboom et al. (1974) who found that subjective color scores were correlated with transmission values, $r = -0.70$. This suggests (from the r^2) that about 50% of the subjective paleness in PSE pork may originate from protein denaturation. Further support comes from Eikelenboom and Nanni Costa (1988) who reported a strong correlation of transmission values with fiber-optic probe measurements, $r = 0.81$.

Myofibrillar Birefringence

Refractive index (n) of various components in meat is given by

$$n = c/v$$

where c is the velocity of light in a vacuum and v is its velocity in various components of meat. Wavelength (λ) decreases with refractive index, and only frequency is constant. In myofibrils, light splits into two components that travel at different velocities, the ordinary ray (O) and the extraordinary ray (E), with O $\perp$ E. Birefringence, which may be either negative or positive in sign is given by $n_E - n_O$. Retardation, the decrease in velocity of light caused by interaction with the medium, may be detected as phase retardation, the interference caused by path difference E $\neq$ O. The path difference of a depth of meat (Γ_m) may be measured by ellipsometry with a de Sénarmont compensator (Pluta, 1988),

$$\Gamma_m \text{ [nm]} = K_\lambda \text{ [nm/degree]} \cdot u°$$

where u is the angle in degrees required for compensation and K_λ is the de Sénarmont constant or path difference for 1° of rotation.

Retardation changes with pH, as shown in Figure 5.1. The path difference always increases as a muscle fiber is taken from the pH of the live animal down to the pH of severely PSE meat (for example, from pH

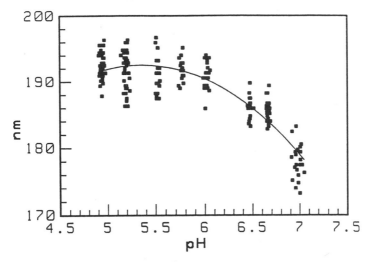

Figure 5.1 *Effect of pH on the optical path difference (nm) of birefringence in a single pork muscle fiber (Swatland, 1989a).*

7.1 to 5.1). The path difference may reach a maximum at this low pH, then start to decrease again if the pH decreases below the isoelectric point. For this technique, single muscle fibers are washed with buffer to remove all optically active components of the sarcoplasm (particularly myoglobin) and are mounted in a chamber under a polarizing microscope. The chamber is flushed between optical measurements by a buffer solution, either controlled manually or by computer (Swatland, 1994). All the measurements are made objectively from a photometer to avoid subjectivity in evaluating the appearance of the microscope image, but the system may be controlled by an operator (Swatland, 1990a) or run completely automatically (Swatland, 1989a). There is some variation in the shape of the relationship of pH with birefringence, because some fibers have a maximum path difference near the isoelectric point of their myofibrillar proteins while others do not, but the general direction of change from pH 7.1 to 5.1 always holds true. Results obtained by polarized light microscopy agree with those obtained with different apparatus, polarized-laser ellipsometry (Yeh et al., 1987).

Results such as those shown in Figure 5.1 represent an average for a whole muscle fiber and are made with the measuring aperture covering the diameter of the whole fiber. Because of the presence of A and I bands along the fiber, however, birefringence is more complex at the ultrastructural level. At the midlength of the A band in a relaxed muscle fiber, the otherwise strong birefringence of the A band is slightly weaker in the H zone, between the ends of thin filaments (Figure 5.2). Similarly, the otherwise weak birefringence of the I band has a slightly stronger region at the midlength of the I band caused by the Z line. Birefringence of the Z line is extremely interesting because it indicates a high degree of protein filament alignment, and we know that weakening and structural alteration of the Z line contributes significantly to the tenderization of beef during conditioning. Increases in birefringence caused by a low pH may increase the light scattering in meat, as seen in the effect of pH on transmittance, perpendicular to the long axes of muscle fibers (Figure 5.3). Thus, as the pH decreases and birefringence increases, less light is transmitted through the fiber, and more light is scattered, relative to a fiber with a high pH and low birefringence.

Relationship to Wavelength and Meat Color

Figure 5.3 also illustrates another optical property of meat, that transmittance of low wavelengths is less than that of high wavelengths. Although meat is a complex optical system, one would still expect to find at least some evidence of Rayleigh scattering (scattering inversely proportional to λ^4) since the lesser dimensions of fibrous muscle proteins are considerably less than the wavelengths of visible light. Meat that has

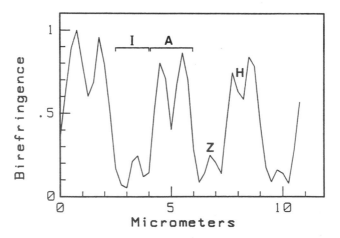

Figure 5.2 *Scanning along the A and I bands of a muscle fiber from the psoas major of a pork carcass thirty hours after slaughter (Swatland, 1989a).*

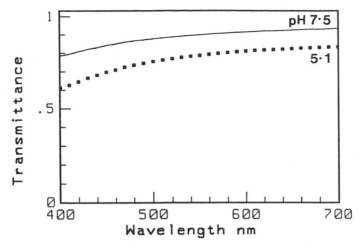

Figure 5.3 *Effect of pH on the transmittance of muscle fibers from the psoas minor of a beef carcass several days after slaughter (Swatland, 1990b).*

been extensively washed to remove its myoglobin and sarcoplasmic proteins has a diffuse surface reflectance spectrum, showing that it acts as a trap for short wavelengths (line 4 in Figure 7.2). Thus, the color of meat seen at its surface is a result of selective absorbance by myoglobin imposed upon a sloping spectrum caused by the scattered light that escapes from meat (as shown by the lower lines of Figure 7.2).

The spectrum of light scattered from meat without chromophores (line 4 in Figure 7.2) may be related back to the transmittance of individual muscle fibers (Figure 5.3), which is related to birefringence (Figure 5.1), ultimately from thick and thin myofilaments (Figure 5.2). There is no contradiction between the contribution of refractive effects to meat paleness and the other two hypotheses (Bendall and Wismer-Pedersen's hypothesis on protein precipitation and Hamm's hypothesis on scattering at the myofibrillar surface). At present, it looks as if refractive effects extend over the whole pH range of meat and that protein denaturation increases light scattering at the lower end of the pH range, depending on temperature. In other words, the refractive effects are reversible, since they depend upon the degree of lateral electrostatic repulsion between myofilaments, and the relationship shown in Figure 5.1 does not depend upon which way the pH is changing. Protein denaturation, however, is irreversible and strongly related to temperature: the combination of acid and high meat temperature increases the extent of protein denaturation.

GONIOPHOTOMETRY

A goniometer is an instrument for measuring geometrical angles, so that a goniophotometer is for measuring light at different angles. Goniophotometry is important for the on-line evaluation of meat because it provides a method to examine critically the scientific basis of PSE and DFD, as well as being adaptable for on-line use. It is important to think in terms of goniophotometry whenever an optical probe is inserted into meat: the angles at which optical fibers and muscle fibers are interfaced is crucial. Goniospectrophotometry is the combination of a goniometer with a spectrophotometer, enabling the measurement of spectra at different angles.

Goniospectrophotometry

Figure 5.4 shows the principle of a fiber-optic goniospectrophotometer for measuring light scattering in meat (Swatland, 1988). Numerous optical fibers are fitted tightly into hypodermic needles inserted in a radial pattern into the meat, like the spokes of a wheel. One optical fiber illuminates the meat at the center of a circle, around which the other optical fibers collect light equidistantly to the tip of the illuminating fiber. The angle of the optical fiber collecting light directly in line with the illuminating fiber is defined as 0°. Thus, an optical fiber perpendicular to both the illuminating fiber and the 0° fiber collects light at 90° to the light passing straight through the meat. The light passing

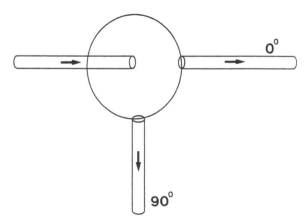

Figure 5.4 *Goniospectrophotometry of meat. One optical fiber illuminates the center of the meat sample while a ring of receiving fibers (only two are shown) collects the light at different angles, equidistantly from the tip of the illuminating fiber.*

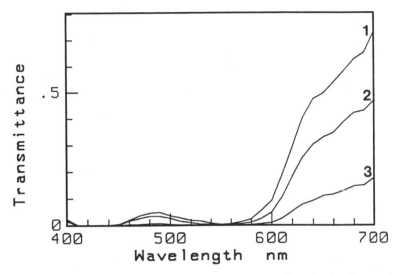

Figure 5.5 *Goniospectrophotometry of pork showing wavelength-related scattering. Spectra 1, 2, and 3 were obtained equidistantly at angles of 0°, 45°, and 90°, respectively. All three angles receive a similar low intensity of light at 400 nm, while light at 700 nm is scattered less and is stronger at 0° than at other angles.*

ward through the meat (out of the illuminating fiber and straight into the 0° fiber) is the strongest relative to other angles, particularly at 700 nm (line 1 in Figure 5.5). Light scattered sideways in the meat is weaker than the straight forward light, particularly at 700 nm (lines 2 and 3 in Figure 5.5); however, at 400 nm, approximately the same intensity of light enters all the optical fibers, from 0 to 90° regardless of angle, because low wavelengths tend to be uniformly scattered through the meat, whereas high wavelengths are scattered less and have a higher forward transmittance.

On-line Laser Scanning

Goniospectrophotometry of meat is a superior method of measuring PSE and DFD but is difficult to adapt for on-line use in a meat plant. A more convenient approach is to scan slices of meat with a laser, sacrificing information on different wavelengths in order to gain enough intensity to penetrate the thickness of a pork chop (Figure 5.6). The laser beam strikes the slice of meat from above, while, below the meat, a photodiode scans from 0° (in line with the laser beam) to 45° to one side of the axis. However, the length of the light path through the meat is no

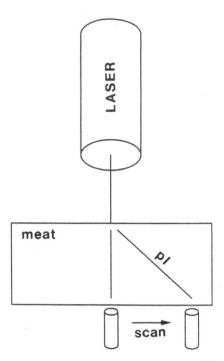

Figure 5.6 *Laser scanning principle of meat introduced by Birth et al. (1978). The laser beam strikes the top of a slice of meat, and the light transmitted through the meat is detected by a photodiode that scans from the optical axis out to one side. The path length (pl) increases as the angle through the meat is increased. In DFD, the laser beam is transmitted with little scattering to give a small, bright spot of light on the underside of the sample, whereas, in PSE, the spot is larger and has a lower intensity because more light has been scattered sideways.*

longer constant (as it was with the fiber-optic goniospectrophotometer described previously).

Laser scanning of meat was introduced by Birth et al. (1978), based on the Kubelka-Munk analysis of light scattering. With the upper surface of a muscle sample illuminated by a helium-neon laser,

$$\log M_T = A - Br$$

where M_T is the radiant exitance on the lower surface, A is the intercept, B is the slope of a regression (light intensity versus distance), and r is the path length through the meat. Birth et al. (1978) showed that

$$B = \log 2 \, (S + K)$$

where *S* is a scatter coefficient (cm^{-1}) and *K* is the absorption coefficient (cm^{-1}). *B* was given no special name by Birth et al. (1978) but may be called a spatial measurement of scattering, for the sake of convenience. Birth et al. (1978) showed that spatial measurements of scattering at 632 nm (the wavelength of a red helium-neon laser) could be used to predict meat quality. Meat is far more complex than the relatively simple situations for which the Kubelka-Munk analysis was intended (Judd and Wyszecki, 1975), and one would expect an additive effect on spatial measurements of scattering by both myofibrillar scattering and chromophore absorbance. The scatter coefficient (*S*) is probably the major variable in spatial measurements of scattering because myoglobin is determined mainly by animal age (at a constant position within a muscle). Despite these extended assumptions, spatial measurements of scattering appear to contain useful information about the physical state of meat.

Given the great improvements that have been made in video image analysis since the 1970s, the method of Birth et al. (1978) could now readily be adapted for sorting pork chops or poultry breasts on-line, using a video camera above meat samples passing over a laser. Obviously, a sample-actuated shutter mechanism would be required to protect the video camera between samples; however, for anyone interested in this method, there are a few subtleties that should be considered.

In the spatial scanning method developed by Birth et al. (1978), sections of pork longissimus dorsi were cut at a thickness of 25 mm, with muscle fibers at an uncontrolled angle (probably about 45°) to the laser beam. Illumination was at 632 nm and the photodetector scanned unilaterally (only to one side of the laser beam) beneath the muscle with a 2-mm diameter measuring aperture. Over a certain range, the logarithm of the photodetector response was approximately linear with respect to the optical path length through the meat, which is how the slope of the intensity-distance relationship could be used as a measure of scattering. But, when this method was tested as a method for sorting turkey breasts on-line, some other factors became apparent (Swatland, 1991).

Following the method of Birth et al. (1978), the length of the light path through the meat was calculated trigonometrically from the position of the photodetector and the depth of the meat sample, and the photodetector response was transformed to a logarithm. The data shown in Figure 5.7 are complete bilateral scans passing completely across the area of meat illuminated by the laser. In a bilateral scan, the magnitude of the hysteresis shows the optical asymmetry caused by muscle fiber arrangement distorting the scattering pattern. In the particular examples shown in Figure 5.7, the scan with the green laser was almost exactly symmetrical while the scan with the red laser shows slight asymmetry. The thickness of each sample is given by the minimum path length.

Both the examples shown in Figure 5.7 are composed of two segments. Nearest to the optical axis, where the light path through the meat is short, there is almost a linear slope in the log response of the photodetector, as expected. Further from the optical axis, however, where the light path through the meat is longer, the angle of the slope decreases, and there is a segment that is almost horizontal, caused by a low level of highly scattered stray light. In the examples shown in Figure 5.7, the junction between these two segments is a transitional curve where attenuated scattering of the incident illumination gives way to the low level of background illumination, and there is no problem in finding the slope at short path lengths. But in data similar to those that might be expected on-line (without precise alignment of muscle fibers or accurate slice thickness), the relationship between photodetector response and optical path length may be curvilinear (Figure 5.8). The lack of a clear separation of the segment nearest to the optical axis presents certain problems in automated computation, which cannot be ignored.

Figure 5.8 also shows an effect that was observed in most samples:

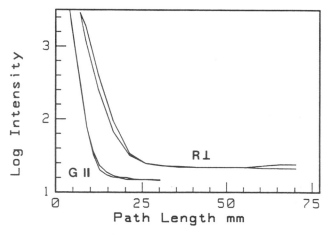

Figure 5.7 *Laser scanning applied to slices of turkey breast meat illuminated by a green laser parallel to, and coaxial with, the longitudinal axes of muscle fibers (G‖) and by a red laser perpendicular to the muscle fibers (R⊥).*

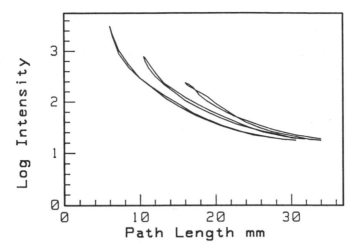

Figure 5.8 *Laser scanning of three slices of turkey breast meat cut at different thickness (given by the minimum path length).*

the relationship between photodetector response and path length was not independent of sample thickness. In the examples shown in Figure 5.8, at a path length of 16 mm, the photodetector response is greater for a thick sample than for a thin sample. In the thick sample at 16 mm, the photodetector is directly in line with the incident illumination, while, in the thin sample at 16 mm, the detector is located to one side of the incident illumination. Thus, there was a higher proportion of transmitted light relative to scattered light for the thick sample at 16 mm than there was for the thin sample at 16 mm. Thus, in practical terms, the method may only be applicable to meat slices of constant thickness, which is a severe practical limitation for sorting raw meat before processing.

The orientation of muscle fibers remains another problem. For the test shown in Figure 5.9, the photodetector was adjusted so that red and green lasers gave similar readings in the optical axis and the depth of the sample was constant. The data obtained by moving the detector across the optical axis were averaged by position (to cancel the hysteresis seen in Figures 5.7 and 5.8). As expected, red light had a greater lateral spread than green light (because of lower absorbance by myoglobin). Lateral spreading was also greater when muscle fibers were perpendicular to the optical axis than when they were parallel and coaxial to the optical axis. Thus, even if sample depth is controlled, attention must be given to the fascicular architecture of the meat. These problems are not insoluble, but neither may they be ignored.

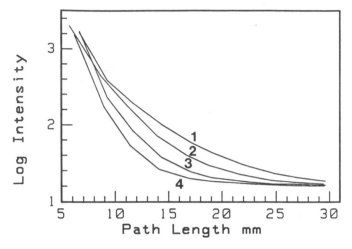

Figure 5.9 *Effect of muscle fiber orientation on laser scanning of turkey breast meat using red (lines 1 and 2) and green (lines 3 and 4) lasers, perpendicular (lines 1 and 3) and parallel (lines 2 and 4) to the longitudinal axes of muscle fibers.*

Wavelength-Position Matrices

Back in the research laboratory, the principle proposed by Birth et al. (1978) also may be used to make goniospectrophotometry more convenient (Swatland and Irie, 1991). It takes several hours to measure a meat sample with a fiber-optic goniospectrophotometer, whereas the same information may be collected in a few minutes by scanning a slice of meat, as shown in Figure 5.6. Both spectrophotometry and spatial measurements of scattering produce a vector of measurements. Thus, if both methods are combined by replacing the laser in Figure 5.6 with an intense source of white light, the product is a matrix with wavelength on one axis and scanning position on the other. Instead of scanning with a simple photodetector beneath the sample, an optical fiber is used to collect light. At each scanning position, the light collected by the fiber is passed through a monochromator to collect a matrix of wavelength-position data. Light is emitted from the lower surface of the sample as a solid angle or steradian and is captured by the area of the distal face of the optical fiber, hence the light entering the fiber is sterance but may be called transmittance, for the sake of familiarity.

The analysis of wavelength-position data uses some of the techniques discussed in Chapter 11 on video image analysis. Scanning positions from −30 mm to +30 mm in steps of 5 mm across the optical axis may be represented by thirteen matrix rows, while wavelengths from 400 nm to 700 nm in steps of 10 nm may be represented by thirty-one columns. Data matrices may be represented pictorially in black and white using pseudo gray-maps, so that each wavelength-position is treated as a pixel (Chapter 11). The probability of a pixel being turned on is proportional to the intensity or gray level at each coordinate (dithering).

Figure 5.10 shows a gray map of the transmittance of white light through two pork samples, PSE and normal. The results were plotted together so that the dithering is the same in both: the probability of a pixel being turned on gives the intensity of the transmitted light. Wavelengths may be read from left to right, 400 to 700 nm. Within the frame for each sample, the position relative to the optical axis may be read vertically, so that the optical axis is represented halfway up each frame. Both samples show that red light at 700 nm is transmitted most strongly, green light weakly, and violet light at 400 nm just slightly. In the normal sample, the transmitted red light is centered on the optical axis, whereas the violet light is spread more or less uniformly across the optical axis from + to − positions. This may be seen more clearly in contour maps of the same data (Figure 5.11). In other words, the wavelength-position matrices in Figure 5.10 show information similar

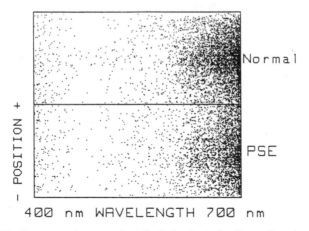

Figure 5.10 *The transmittance of white light through slices of pork, scanned by apparatus similar to that in Figure 5.6, but illuminated with white light instead of a laser, and scanning with an optical fiber leading to a spectrophotometer instead of scanning with a photodiode. Slices from normal and PSE pork have been plotted together to use the same probability of a laserjet pixel being turned on in proportion to the intensity of transmitted light.*

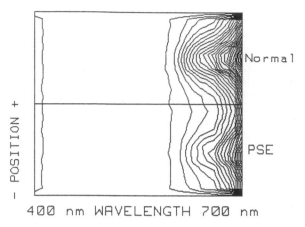

Figure 5.11 *The same data as in Figure 5.10 plotted to show contours of equal light intensity.*

to that shown by goniospectrophotometry in Figure 5.5 and by laser scanning in Figure 5.8.

Why bother with such a complex procedure? There are three reasons. First, in measuring two things at once (wavelength and position), the retrieval of information from the sample has been increased: what may be missed by one method might be captured by the other. Second, we may use the full power of photodiode arrays and video cameras to capture a matrix instantaneously. Third, there is now more information for multivariate analysis, as shown in Table 5.1. Note how often the low wavelengths are selected by stepwise regression as the most useful predictors.

In Table 5.1, where an important correlation of transmittance with a meat quality attribute is relatively weak, it may be improved by using the other information collected at a different scanning position. Thus, for the correlation of optical properties with slicing loss at thawing, which is important in preparing thin slices of pork for Japanese cuisine, the correlation may be increased from $R = 0.70$ to 0.82, using information from three scanning positions off the main axis (Swatland and Irie, 1991).

As a generalization, spectrophotometry straight through a slice of meat contains more reliable information about PSE than monochromatic measurements made at different angles through the meat, and, in the latter category, wavelengths from 610 to 700 nm are more useful than those from 400 to 600 (Swatland and Irie, 1992). From a practical perspective, correlations are not much use in predicting meat quality until the r^2 or R^2 is over about 0.9, so we cannot afford to waste any information that might enable us to improve our prediction equations. In the laboratory, using stepwise regression of wavelength-position matrices, it is possible to reach $R = 1$ for the prediction of PSE (Swatland and Irie, 1992), but it is doubtful whether this could be done industrially for a large population. It is important, however, to demonstrate that something is possible, even if only in the laboratory. Laboratory results achieved under ideal conditions give us an achievement target for on-line applications in the real world.

The information content of wavelength-position matrices also may be revealed by plotting simple correlation coefficients instead of light intensity. Figure 5.12 shows a contour map for negative and positive correlations of transmittance with the subjective Japanese pork color scores of Nakai et al. (1975). The presence of correlations near to zero at the extremes of the range for position (top and bottom rows of each matrix in Figure 5.12) showed that the range (from -30 to $+30$ mm relative to the optical axis in this case) was large enough to catch all the useful information. However, one would expect this range to change with the thickness and nature of the samples.

Table 5.1 *Multiple Correlations (P < 0.01) of Pork Quality Attributes with Transmittance at Different Wavelengths (only the three most useful wavelengths are shown; Swatland and Irie, 1991).*

Quality Attribute	R	Wavelength (nm)		
Drip loss before freezing	0.70	400	610	
Drip loss after freezing and thawing	0.71	700	450	590
Slicing loss before freezing	0.55	430	450	470
Slicing loss at thawing	0.70	700	590	470
Japanese pork color score, unfrozen	0.75	690	490	590
Japanese pork color score, after freezing and thawing	0.82	610	590	490
Japanese pork color score, four days after freezing and thawing	0.92	490	590	560
Water holding capacity, unfrozen	0.78	470	490	510
Water holding capacity, after freezing and thawing	0.56	640	400	580

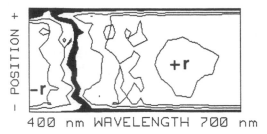

Figure 5.12 *Simple correlations of the transmittance of white light through pork slices with subjective Japanese pork color scores plotted by their wavelength-position coordinates* ($r = -0.64$ to 0.77, $P < 0.005$). *In the contour map, the interval is* $r = 0.20$; *the black area in the center shows a region with* $r = 0$, *separating regions that are negatively* ($-r$) *and positively correlated* ($+r$) *with the color scores.*

In Figure 5.12, correlations are positive above about 470 nm because the Japanese pork color scores are low for PSE and high for DFD. Thus, dark pork with a high color score transmits more red light than PSE pork with a low color score. This agrees with the method proposed by Birth et al. (1978) using a red laser. However, something not seen with the monochromatic laser is that the opposite condition occurs at low wavelengths: relative to dark samples, PSE pork with a low color score has higher transmittance at wavelengths <470 nm. Thus, almost the whole wavelength-position matrix contains useful information, not just the red peaks as seen with laser scanning.

SURFACE AND INTERFACE ANISOTROPY

We have already discussed the birefringence of meat caused by the ultrastructural alignment of myofilaments, but bulk meat also exhibits optical anisotropy at the cellular and fascicular level of organization. This form of optical anisotropy is detected when reflectance is measured carefully with a colorimeter or fiber-optic spectrophotometer. With an integrating sphere colorimeter, more light is reflected when muscle fibers are parallel to the sample surface than when they are perpendicular (looking ahead to the chapter on meat color, compare line 1 with 3, and 2 with 4 in Figure 7.4). With fiber optics, reflectance from meat is higher when the interface between optical fibers and muscle fibers is perpendicular than when the interface is coaxial (compare spectra 3 and 4 in Figure 7.7).

Two factors may contribute to optical anisotropy in bulk meat. First, muscle fibers or myofibrils themselves may act as optical fibers, tending to conduct light along their length by a series of internal reflections. Second, the longer side profiles of thick and thin myofilaments may lead to greater scattering than their small end profiles; however, this important optical property of bulk meat has not yet been investigated experimentally, and we cannot be sure of these hypotheses.

Surface Reflectance Spectrum

Relative to the visual appearance of meat, the optical properties of meat involve several levels of organization:

(*1*) Sliding filaments and myofibrillar birefringence
(*2*) Myoglobin and protein precipitation in the sarcoplasm
(*3*) Postmortem fiber shrinkage and the release of fluid
(*4*) The macroscopic surface reflectance properties of cut meat

Cut-surface properties are extremely important because they determine what the meat looks like on the display counter, but our probes and research techniques are reading from the lower levels of organization. It is extremely important, therefore, to understand how levels 1, 2, and 3 create the properties at level 4. But this is yet another of those areas where we have not reached a scientific consensus, and the reader needs a critical approach to what follows. It is possible to assemble reasonable hypotheses to explain the factors underlying the visual appearance of meat, but most unfortunate that we cannot then provide extensive citations to critical experiments testing these hypotheses.

When a cut surface of meat is illuminated with white light, some light is reflected directly from the wet surface as specular reflectance with no change in wavelength. Of the light that enters the meat, most is scattered, and its original direction gradually is lost as it penetrates deeper into the meat. A fraction of this scattered light makes a U-turn after being scattered at a myriad of points and returns from whence it came. Some eventually escapes from the meat surface and is visible to the observer: the more light that is returned across the meat surface, the more pale the meat appears to be, and vice versa in the case of DFD. While passing through the meat, some wavelengths are absorbed: most of the green light is absorbed by myoglobin. Thus, meat with a high myoglobin content (such as beef) appears to be red to the observer, although meat with a very low myoglobin content (such as turkey breast meat) may appear only slightly pink. If a high degree of scattering shortens the light path through the meat, the opportunity for selective absorbance by myoglobin is decreased, and the meat appears less red or pink than normal (as well as being more pale than normal). In combination, these two factors may cause the almost white appearance of severe PSE pork—when the observer must look closely to find the boundary between muscle and fat. Metmyoglobin formation occurs most rapidly in subsurface layers of the meat, so there may be a hidden brown filter beneath the surface. If this occurs, the initial redness or pinkness of the meat appears to fade away gradually, until the brown metmyoglobin layer reaches the surface, at which point the meat appears brown.

The internal reflectance spectrum obtained from levels 1, 2, and 3 by a fiber-optic probe inserted into the meat may be flat relative to a reflectance spectrum measured above the cut surface of the meat, but it may be transformed by the cubic power of wavelength to resemble a surface reflectance spectrum (Figure 5.13). The curvilinear shape of the conventional surface reflectance spectrum is caused by the selective absorbance of myoglobin superimposed on a skewed, but otherwise

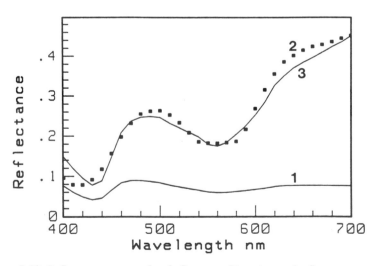

Figure 5.13 *Reflectance spectra of pork. Spectrum 1 is an internal reflectance spectrum obtained directly from the meat via optical fibers, spectrum 2 (dotted) is a conventional surface reflectance spectrum obtained at 45° from the meat surface, and spectrum 3 is the internal reflectance spectrum transformed by the third power of wavelength (Swatland, 1989b).*

relatively flat, spectrum (line 4 in Figure 7.2) that is created by the meat trapping light at low wavelengths.

Fairly strict protocols are used when measuring surface reflectance spectra (Judd and Wyszecki, 1975), and we tend to forget how variable surface reflectance spectra may be if we change the measuring protocol. With a probe actually within the meat, the measuring conditions are radically different to those that prevail at the surface. In particular, there is an air space between the optical components and the meat when a surface reflectance is measured, but direct contact between optical components and the meat when an internal reflectance is measured. Other sources of variation in internal reflectance spectra, often compounded with the inherent optical anisotropy of meat, originate from the optical geometry of the probe window, as discussed in Chapter 3.

SARCOMERE LENGTH

Laser Diffractometry

Cold-shortened meat with short sarcomeres is tough, while stretched meat with long sarcomeres is tender, so that measuring sarcomere length is an important analytical method in the meat laboratory. Good results may be obtained by light microscopy, but probably it is easier to make large numbers of measurements using a simple laser diffractometer (Rome, 1967), as shown in Figure 5.14. Both microscopy and diffractometry require careful dissection and preparation of the muscle, but the latter does not require sample homogenization, and data may be collected with a simple calliper measurement, rather than with a microscope eyepiece micrometer.

$$\text{Sarcomere length} = \lambda/\sin\Theta$$

where Θ is the angle subtended by the first-order diffraction band relative to the optical axis. Because the angles are relatively small, $\sin\Theta \approx \tan\Theta$, given by S/D in Figure 5.14. Diffractometry requires the dissection of a single bundle or part of a bundle of muscle fibers, but it may be impossible to obtain a clear definition of the diffraction band if the fibers are kinked, as often happens when some fibers cold-shorten while others do not.

It has not yet been possible to apply diffractometry for the measurement of sarcomere lengths on-line in intact carcasses, although this would be extremely useful in detecting tough beef caused by cold-shortening. The main problem is that diffraction patterns are rapidly lost by scattering in bulk meat.

NIR Birefringence

Although no progress has been made in on-line diffractometry, an attempt has been made to use birefringence measurements. With two plane polarizers (a fixed polarizer and rotatable analyzer), the maximum transmittance occurs when the analyzer is rotated parallel to the polarizer, and the minimum happens when the analyzer is perpendicular to the polarizer. But if the polarizer is at $0°$ when the analyzer is at $90°$, then a birefringent muscle fiber at $45°$ between the polarizer and analyzer rotates light so that now it may pass through the analyzer. As we saw earlier in Figure 5.2, the brightest bands are the A bands, where thick (myosin) and thin (actin) filaments overlap. But maximum transmittance now is at an analyzer angle $> 90°$, because of the optical path difference of the muscle fiber.

The apparatus shown in Figure 5.15 uses NIR to minimize scattering and enables measurements to be made on a hand-cut slice of meat held between two glass plates to control sample thickness. Light from a halogen projector bulb passes through an interference filter (800 nm) and into a fiber-optic light guide acting as a depolarizer, through a polarizer and a 9-mm aperture, and through the sample. The transmitted light then passes through an analyzer to a photodiode.

The data shown here have $0°$ as the position where the analyzer is perpendicular to the polarizer (an offset of $90°$ from usual methods of presentation), the advantage being that, for a slice of meat, this places the peak of the sine wave near the center of the graph. Thus, for a

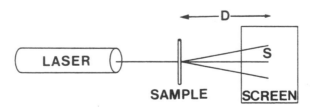

Figure 5.14 Laser diffractometry for the measurement of sarcomere length using the distance (D) between the sample and the screen, and the separation (S) of the diffraction band from the optical axis.

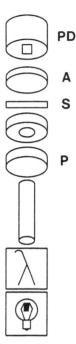

Figure 5.15 *Apparatus to measure NIR birefringence in a slice of meat (S) using a polarizer (P) and analyzer (A).*

standardization curve measured without a meat sample in position, when the analyzer is rotated to 90°, transmittance, $T = 1$. This follows the Malus Law for crossed polarizers (Pluta, 1988), where T for the azimuth angle (α) between the axes of the polarizers (= offset + abscissa) is given by

$$T_\alpha = T_{90} + (T_0 - T_{90}) \cos^2\alpha$$

Thus, if a sample with high scattering but no birefringence (such as ground glass) is measured, the peak transmittance is still at 90° (although < 1). When slices of meat are measured, with high scattering plus birefringence, the birefringence is detectable because peak transmittance is at an angle $> 90°$. Research is still in progress, and we have not yet been able to use the angle of rotation properly (as was done with a polarizing microscope to obtain Figure 5.1). From Kerr (1977), one would expect the phase difference (P) to be determined by birefringence ($n_2 - n_1$), sample thickness (t), and wavelength (λ),

$$P = t \cdot (n_2 - n_1) \cdot \lambda^{-1}$$

With phase differences from 15° to 25° that occur in a 1-mm slice of meat, one would expect birefringence from 0.02 to 0.012. But when the interference colors with white light are examined, they are not uniform across the whole specimen, tending to form a mosaic and creating a serious sampling error. Thus, it is difficult or impossible for ellipsometry.

Despite these problems, however, even the simple transmittance of NIR polarized light is affected by sarcomere length. Transmittance increases as sarcomere length increases from 1.2 to 1.5 m; it peaks at sarcomere length 1.5 μm, then decreases as sarcomere length increases to 3.5 μm, as shown in Figure 5.16. Figure 5.2 shows that birefringence is strongest where thick and thin filaments overlap. Thus, if bulk birefringence also is proportional to the overlap of thick and thin filaments, it will tend to increase as sarcomere length is decreased, but, in cold-shortened meat where thin filaments start to overlap, there may be a loss of alignment that disrupts the birefringence. Thus, in restrained muscles without cold-shortening, the transmittance of polarized NIR may enable the prediction of sarcomere length, but the relationship is lost when cold-shortening occurs and myofibrillar birefringence is disrupted (Figure 5.17). As Figures 5.16 and 5.17 show, the correlation of transmitted NIR with sarcomere length in meat slices is lost if data from short and long sarcomeres are pooled (since $r < 0$ with sarcomere length < 1.5 μm and $r > 0$ with sarcomere length > 1.5 μm).

To summarize this method, although sarcomere length is detectable in

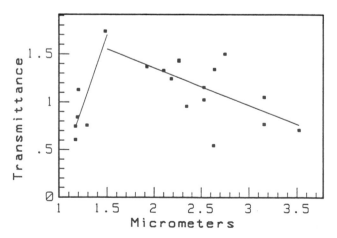

Figure 5.16 *Transmittance of polarized NIR in relation to sarcomere length in pork slices.*

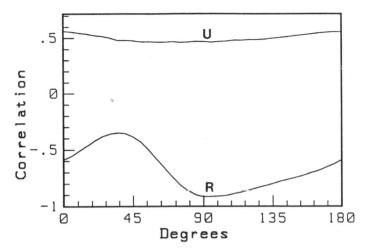

Figure 5.17 *Transmittance of polarized NIR through restrained pork with relatively long sarcomeres (R, mean sarcomere length 2.52 ± 0.63 µm) and unrestrained, cold-shortened pork (U, mean sarcomere length 1.82 ± 0.67 µm), showing the correlation of transmittance with sarcomere length.*

bulk slices of meat, stretched sarcomeres and highly cold-shortened meat both have low transmittance of polarized NIR, which is a difficult problem to circumvent in adapting this technique for on-line use. Another major problem is that the technique is exquisitely sensitive to pH. Perhaps this approach could take us a step nearer to an on-line probe for sarcomere length, because at least it uses bulk meat slices, rather than dissected fibers, but still we have many more steps to go. Some data for a meat processing application are given in Chapter 12.

REFRACTOMETRY

Measurement of the refractive indices of meat fluids is relatively straightforward, using an Abbe refractometer; however, anyone attempting to use an existing on-line refractometer borrowed from a chromatography laboratory should first check its linearity, since some instruments are designed for maximum sensitivity at the start of a short, sharp refractive index peak and may not give a reliable reading of a slowly changing or sustained refractive index. The refractive indices of the fluids lost from meat (drip and exudate) are extremely variable, as seen in Table 5.2.

One might expect the refractive index of meat drip to decline with time, arguing that the first fluid released should be sarcoplasm, with a

Table 5.2 *Refractive Indices (n) of Drip Losses from Pork, Showing (n − 1.35) × 10,000 [from Irie and Swatland (1992)].*

Temperature	Day on Test		
	1	2	3
5°C	83.4 ± 40.0	89.9 ± 42.3	56.5 ± 24.0
22°C	86.4 ± 41.5	99.2 ± 40.9	

high content of soluble protein, while the final fluid released should be from within the myofilament lattice, low in soluble protein; however, this has not yet been reported in practice. If anything, the effect is reversed, at least initially. Despite this uncertainty, the refractive index of drip from pork slices is strongly correlated with PSE ($r = 0.94$ for the correlation of refractive index with reflectance at 700 nm; Irie and Swatland, 1992). Another use of the data in Table 5.2 is to help in the design of fiber-optic probes, where the refractive index of the fluid in contact with optical fibers has a major effect in determining the interface properties, as considered earlier in Chapter 3.

The angle of the cone of light emitted from an optical fiber depends on the relationship of the refractive indices of the core versus the cladding. After stripping off the cladding from a quartz optical fiber and

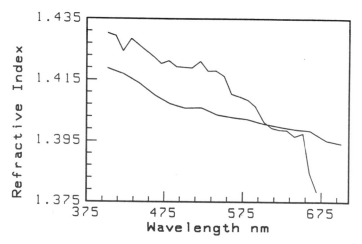

Figure 5.18 *Refractive index of two samples of beef in relation to wavelength (Bolin et al., 1989).*

wrapping it with meat, Bolin et al. (1989) were able to measure the refractive index of beef from

$$N_s = \text{SQR} \{N_q^2 - [n_0 \sin (\Theta)]^2\}$$

where Θ is the half angle of the emergent cone, n_q is the refractive index of the optical fiber, n_s is the refractive index of the tissue, and n_0 is the refractive index of air. Beef from sixteen animals was measured, and the refractive index was quite variable, 1.412 ± 0.006. The effect of wavelength on two samples of beef is shown in Figure 5.18, where, as expected, refractive index declines with wavelength. It may be possible to measure the cone angle with a photodiode array, thus making it possible to measure the refractive index of meat on-line.

REFERENCES

Bendall, J. R. 1962. *Proceedings of the First International Congress of Food Science and Technology.* London, J. M. Leitch, ed., New York: Gordon and Breach Science Publishers, pp. 23−30.

Bendall, J. R. and H. J. Swatland. 1988. *Meat Sci.,* 24:85.

Bendall, J. R. and J. Wismer-Pedersen. 1962. *J. Food Sci.,* 27:144.

Birth, G. S., C. E. Davis, and W. E. Townsend. 1978. *J. Anim. Sci.,* 46:639.

Bolin, F. P., L. E. Preuss, R. C. Taylor, and R. J. Ference. 1989. *Applied Optics,* 28:2297.

Dekker, T. P. and H. G. Hulshof. 1971. *Proceedings of the 2nd International Symposium on Condition and Meat Quality of Pigs.* Zeist, March 22−24, 1971, Wageningen: Centre for Agricultural Publishing and Documentation, pp. 79−80.

Eikelenboom, G. and L. Nanni Costa. 1988. *Meat Sci.,* 23:9.

Eikelenboom, G., D. R. Campion, R. G. Kauffman, and R. G. Cassens. 1974. *J. Anim. Sci.,* 39:303.

Engelmann, T. W. 1878. *Pflügers Arch. Ges. Physiol.,* 18:1.

Hamm, R. 1960. *Advances Food Res.,* 10:355.

Hart, P. C. 1962. *Tijdschr. Diergeneesk.,* 87:156.

Hartshorne, N. H. and A. Stuart. 1970. *Crystals and the Polarising Microscope.* 4th edit., London: Edward Arnold, pp. 309−677.

Irie, M. and H. J. Swatland. 1992. *Food Res. Internat.,* 25:21.

Judd, D. B. and G. Wyszecki, G. 1975. *Color in Business, Science and Industry.* New York: John Wiley & Sons, pp. 420−431.

Kauffman, R. G., G. Eikelenboom, P. G. van der Wal, B. Engel, and M. Zaar. 1986. *Meat Sci.,* 18:307.

Kauffman, R. G., R. G. Cassens, and D. L. Meeker. 1993. *Meat Internat.,* 3(2):40.

Kerr, P. F. 1977. *Optical Mineralogy.* 4th edit., New York: McGraw-Hill, pp. 92−99.

Lawrie, R. A., D. P. Gatherum, and H. P. Hale. 1958. *Nature,* 182:807.

Martin, A. H. and H. T. Fredeen. 1974. *Can. J. Anim. Sci.,* 54:137.

Nakai, H., F. Saito, T. Ikeda, S. Ando, and A. Komatsu. 1975. *Bull. Nat. Inst. Anim. Indust., Chiba, Japan,* 29:69.

Offer, G. and J. Trinick. 1983. *Meat Sci.,* 8:245.

Offer, G., P. Knight, R. Jeacocke, R. Almond, T. Cousins, J. Elsey, N. Parsons, A. Sharp, R. Starr, and P. Purslow. 1989. *Food Microstruct.,* 8:151.

Pluta, M. 1988. *Advanced Light Microscopy, Vol. 1.* Amsterdam: Elsevier, p. 38.

Rayment, I., W. R. Rypniewski, K. Schmidt-Base, R. Smith, D. R. Tomchick, M. M. Benning, D. A. Winkelmann, G. Wesenberg, and H. M. Holden. 1993. *Science,* 261:50.

Rome, E. 1967. *J. Mol. Biol.,* 27:591.

Schmidt, G. R., L. Zuidam and W. Sybesma. 1971. *Proceedings of the 2nd International Symposium on Condition and Meat Quality of Pigs.* Zeist, March 22–24, 1971, Wageningen: Centre for Agricultural Publishing and Documentation, pp. 73–79.

Stanley, D. W., A. P. Stone, and H. O. Hultin. 1994. *J. Agric. Food Chem.,* 42:863.

Swatland, H. J. 1988. *J. Anim. Sci.,* 66:2578.

Swatland, H. J. 1989a. *J. Comput. Assist. Microsc.,* 1:249.

Swatland, H. J. 1989b. *Can. Inst. Food Sci. Technol. J.,* 22:165.

Swatland, H. J. 1990a. *Trans. Amer. Microsc. Soc.,* 109:361.

Swatland, H. J. 1990b. *J. Anim. Sci.,* 68:1284.

Swatland, H. J. 1991. *Can. Inst. Food Sci. Technol. J.,* 24:27.

Swatland, H. J. 1994. *J. Comput. Assist. Microsc.,* 6:41.

Swatland, H. J. and M. Irie. 1991. *J. Comput. Assist. Microsc.,* 3:149.

Swatland, H. J. and M. Irie. 1992. *J. Anim. Sci.,* 70:2138.

Yeh, Y., R. J. Baskin, K. Burton and J. S. Chen. 1987. *Biophys. J.,* 51:439.

CHAPTER 6

Electrical Properties of Meat

INTRODUCTION

THE LIST OF those who have contributed to our knowledge of the electrical properties of skeletal muscle includes some of the greatest scientists who have ever lived, from the paramount discoveries of Galvani to the almost forgotten studies of Faraday on the magnetic properties of muscle. Muscle has high conductivity (low resistance), while fat has low conductivity (high resistance), because muscle contains a continuous labyrinth of electrolytes while fat is dominated by spheres of lipid. This enables the boundary between fat and muscle to be detected from the difference in resistance and enables the overall lipid content of a volume of meat to be detected electromagnetically. Within the muscle, the membranes partitioning the labyrinth of electrolytes are changed during the development of rigor mortis, enabling the detection of PSE from resistance and capacitance.

No matter how we start evaluating meat on-line, we always conclude by processing our data electrically. Thus, techniques that relate directly to the electrical properties of meat may have an immediate advantage, as well as being the beneficiary of future progress in electronic instrumentation.

BASIC ELECTRICAL PROPERTIES OF MEAT

Cellular Components

Meat is derived from a living tissue—striated skeletal muscle. Although meat is more or less a dead tissue (some enzymes still may be active in cured meat), the earliest on-line evaluations of meat after slaughter are potentially the most useful, so understanding the dynamic transition from living muscle to meat is important. At present, many types of measurements made soon after slaughter are unreliable, for one reason or another, and this prevents us from taking advantage of possibilities for differential refrigeration protocols. Measurements soon after slaughter are much easier to relate back to the animal and farm origin, because ear tags, skin tattoos, or transponders are still close at hand.

The published literature on the electrical properties of living muscle is quite voluminous, representing a century or more of dedicated effort by ever-increasing numbers of medical researchers, biophysicists, and physiologists. All that is attempted here is a simple account of a few basic electrical properties of meat.

Figure 6.1 introduces the main components. Meat is composed of muscle fibers, which are very large cells often many centimeters in length. Like all cells, muscle fibers are bounded by a plasma membrane with strong dielectric properties and a capacitance of approximately 1 μF cm^{-2} (Cole, 1970). As expounded by Catton (1957), the cell membrane is a very effective insulator, which, if scaled up from microscopic to macroscopic dimensions, would be separating fluids at about 80 kV cm^{-1}, while maintaining perfect flexibility. Thus, if gross

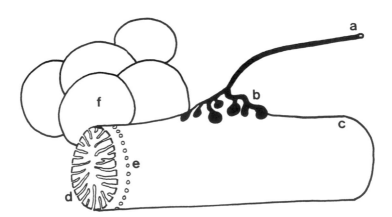

Figure 6.1 *Key components in explaining the electrical properties of meat. A terminal axon (a) innervates a neuromuscular junction (b) on the surface of a muscle fiber (c). Only a short length of the muscle fiber is shown. On the cut surface of the fiber, transverse tubules are shown radiating inwards to increase the membrane area of the fiber (d). Transverse tubules open on the surface of the muscle fiber in rings, two per sarcomere at the A-I junction, but only one ring is shown (e). Globular fat cells (f) may occur anywhere between fibers, between bundles of fibers, or between muscles.*

electrodes (large electrodes relative to the muscle fibers) are inserted into the meat of a recently slaughtered carcass, strong overall capacitance is detectable. Before exploring this important observation any further, there are several other points to make from Figure 6.1.

As discovered by Faraday (1845), a muscle fiber is more diamagnetic perpendicular to its long axis than parallel to its long axis (Arnold et al., 1958). This magnetic anisotropy is unaltered by freezing and thawing but changes in sign when freeze-dried muscle is rehydrated. Diamagnetism is lost as muscle fibers are dried, and, eventually, they may become paramagnetic.

Meat contains a variable amount of fat: under the skin (subcutaneous fat), between the muscles (intermuscular fat), within the muscles (intramuscular or marbling fat), around organs in the abdominal cavity (visceral fat), cushioning at critical parts of the skeleton (structural fat), within marrow cavities of bones from older animals, and so on. The locations of fat deposits have a high degree of anatomical consistency, and there are well known growth patterns for different fat deposits in the meat animal carcass. Figure 6.1 shows some representative adipose cells next to the muscle fiber. The key point is that, like the lipid membrane, the large triglyceride droplet in the adipose cell also acts as an insulator and contributes capacitance to the circuit of a test current through the meat.

Another point to be gleaned from Figure 6.1 is that muscle fibers are innervated. A terminal axon from a motor neuron in the ventral horn of the spinal cord (one of hundreds or thousands of axons from the same neuron) makes contact with the muscle fiber at a neuromuscular junction. Transmission of impulses along both axons and muscle fiber membranes is by a rapid, self-propagating flux of sodium and potassium ions known as the action potential, although to pass across the neuromuscular junction requires the release, diffusion, and receipt of a neurotransmitter molecule, acetylcholine. During the time between action potentials (in other words, when the activation-contraction system is quiet), the membranes of axons and muscle fibers restore a resting potential in the order of -84 mV from outside to inside the cell. Some of this potential is a Donnan equilibrium of ions across the membrane, but the remainder originates from the ATP-powered sodium ion pump of the membrane. Muscle innervation is important for on-line meat evaluation for several reasons.

The nervous system is far more sensitive to stimulation than is muscle, so that a variety of extrinsic factors (reflex neural activity, noise from cut axons, and postmortem electrical stimulation) all reach the muscle and may cause contraction via the nervous system. In on-line testing of pork soon after slaughter, it would be extremely undesirable to accelerate glycolysis and decrease the pH by applying an on-line test current anywhere near the spinal cord, because this might exacerbate the development of PSE. Thus, the nervous system is the arrival pathway for many environmental factors affecting meat quality. Plate 20 gives a sketch of the dynamic system we are dealing with, and the full story is given elsewhere (Swatland, 1994).

A final point to glean from Figure 6.1 is that the muscle fiber internally activates its thousands of myofibrils by conducting action potentials down countless inpushings of the surface membrane, each of which is called a transverse tubule. The fluid of the extracellular space between the muscle fibers is continuous with the fluid-filled transverse tubules running into the depth of each muscle fiber, thereby creating a dramatic increase in the membrane area responsible for the overall capacitance of meat. Within the muscle fiber, there are other very large membrane areas, such as the sarcoplasmic reticulum. Whether or not the sarcoplasmic reticulum contributes to the gross electrical capacitance of meat is an open question.

Resting Potentials

If a muscle with a relatively slow metabolism, such as beef sternomandibularis, is removed after slaughter and impaled with a glass microelectrode, the concentrated potassium chloride solution in the axis of the glass microelectrode detects resting potentials in the muscle fibers (Figure 6.2). Resting potentials may persist for hours after slaughter, showing that not only does the membrane maintain its integrity for some time, but also that it has a supply of energy.

Not surprisingly, resting potentials in meat are never as stable or reliable as those in living muscle, but the fact that they persist for so long is quite remarkable. The response to stimulation (irritability) of groups of muscle fibers innervated by the same motor neuron (a motor unit) is increased by surface drying and aerobic exposure of the meat but inhibited if the surface is wet, anaerobic, or cold. Even though axons may have been cut to remove a sample for measurement, transected axons often are resealed within several minutes by membrane lipids, depending on temperature and the extracellular concentration of calcium

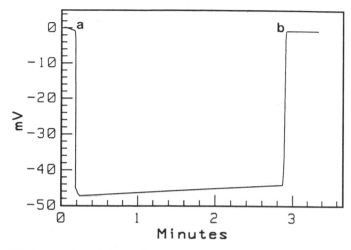

Figure 6.2 *Resting potential recorded from a bovine sternomandibularis muscle fiber several hours after slaughter as the glass microelectrode pushed through the cell membrane (a) and then was withdrawn (b) [From Swatland (1980c)].*

ions acting on phospholipase (Yawo and Kuno, 1983). A transected axon on the meat surface that has been resealed may initiate action potentials in the remaining parts of the axon in the meat, thus leading to muscle fiber contraction. Even if the muscle fiber is transected (to get it into the apparatus), it may be plugged by a diffusion barrier derived from membrane phospholipids and dependent on calcium ions for its formation (Echeverria et al., 1983). After a fiber in an experiment has been sealed, the inward diffusion of dyes from the extracellular space is halted and the resting potential of the fiber may be restored.

Having used the existence of cellular resting potentials in beef as evidence that the gross electrical capacitance of meat originates from cell membranes (Guttman, 1939), it is important to balance the picture by stating that resting potentials in the meat from other species may be lost very rapidly after slaughter. Even the persistence of resting potentials in beef may owe much to the survival of aerobic metabolism on the surface of the sample, where the yield of ATP from the glucose units of stored glycogen will be far higher than in the interior of the samples (Pasteur effect). Also, even in living muscle, the accumulation of lactate is associated with decreases in membrane resting potentials (Jennische et al., 1978). Thus, the loss of muscle fiber resting potentials is quite rapid in pork, particularly if it is PSE (Schmidt et al., 1972). Schmidt et al. (1972) found resting potentials in pork forty-five minutes after slaughter that ranged from 21 to 48 mV in carcasses from PSS pigs and from 38 to 62 mV for normal pigs. This explains how it is possible to use impedance or capacitance for the on-line evaluation of PSE. Conductivity may be used to determine the age of a bruise on a carcass (Hamdy et al., 1957), but this lead has never been followed industrially.

Polarization

It is a great advantage to have at least some type of model for the electrical properties of meat. It encourages the scientist to look critically at the causes of phenomena and provides an interface that is familiar to collaborators with a background in electronics. Meat is a highly reactive mixture of biochemicals and, if impaled by metal electrodes, readily produces a range of galvanic effects, which, of course, is how the physics of electricity and physiology of nerves and muscles both got started in the laboratory of Luigi Galvani more than 200 years ago. If a DC current is applied to a couple of stainless steel shrouding pins in a beef carcass and then disconnected and replaced by a flashlight bulb, the beef carcass may keep the bulb alight for quite a while if it does not immediately burn it out. Thus, when we attempt to measure the electrical properties of meat, the first concern is with electrode polarization. The accumulation of hydrogen bubbles on the negative plate of a galvanic battery impedes the output of electricity, and comparable problems exist with a DC current in meat if we attempt to measure the resistance of the meat.

When a metal electrode is in contact with electrolytes in meat (fluids within and between muscle fibers), the electrode tends to discharge ions into the fluid, and ions in the fluid tend to combine with the electrode. This may create a charge gradient arranged as an electrical double layer, and the electrode-electrolyte interface may behave as a voltage source and a capacitor in parallel with a resistor. Each of the following has an effect on this system:

(*1*) The metallic composition of the electrode

(*2*) The area and shape of the electrode in contact with meat electrolytes

(*3*) The variable blocking of parts of the electrode area by insulation from adipose cells

(*4*) Current density and the separation and arrangement of electrodes

(*5*) Changes in electrolyte composition caused by progressive changes in the postmortem metabolism of meat, such as an increase in acidity causing the release of fluid with a low concentration of electrolytes from the myofilament matrix

(6) The temperature of the system, allowing that the electrodes may require an equilibration time to reach the temperature of a hot carcass or of a refrigerated carcass and that carcasses may be at different temperatures even at a set time after slaughter, because some have more exothermic glycolysis than others, while some have more adipose insulation than others

(7) The frequency of the test current. The resistance and capacitance in parallel of meat are inversely proportional to the logarithm of frequency (Swatland, 1986).

The difference in potential between identical electrodes would not be a problem if it was constant, but it is not. In the medical field, many of these problems may be avoided by using electrode pastes or by depositing a fragile layer of chloride on electrodes, but this is not feasible in the meat industry for any electrode that is to penetrate the meat. Pastes contaminate the meat, while friction removes the chloride. The answer for on-line meat evaluation is either to use an AC current if two electrodes are used (Figure 6.3), or to use four electrodes with two supplying the test current and two detecting it, either for DC or AC (Burger and van

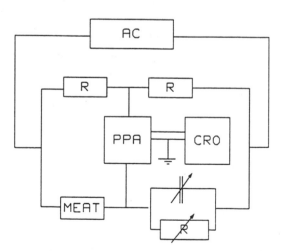

Figure 6.3 *Two-electrode Wheatstone bridge circuit of Burger and van Dongen (1960), adapted for an isolated meat sample. This would not be suitable for a hanging carcass because of the grounded circuit. AC power source, AC: resistors, R; push-pull amplifier, PPA; and cathode-ray oscilloscope, CRO.*

Dongen, 1960). The Wheatstone bridge for two electrodes may be altered in any way that is convenient and is ideal for a dual trace oscilloscope (Swatland, 1980a).

Frequency

The frequency dependency of the dielectric constant of muscle overall is an inverse relationship but has three zones of nonlinearity or relaxation regions (Schwan and Kay, 1957):

- α: Around 100 Hz, there is an interaction of the test current with excitable cell membranes and large intracellular spaces.
- β: Around 10^6 Hz, charging and discharging of a heterogenous mixture of dielectrics in the cell occurs (Maxwell-Wagner effect).
- γ: Around 10 GHz, there is rotation of electric dipoles (Debye effect) in water and protein.

Numerous models have been proposed for the gross electrical properties of meat, some of them quite inappropriate and missing the point that meat is a highly anisotropic system composed of large numbers of insulating cylinders in various states of decay (muscle fiber membranes) enclosing and being surrounded by sometimes separate electrolytes (intra- and extracellular fluid, respectively). The electrical anisotropy of muscle has been known for well over a century (Epstein and Foster, 1983). But even models that take this inherent anisotropy into account should not omit that muscle fibers in meat often are not parallel but have an angular arrangement (pennate structure). The model shown in Figure 6.4 attempts to explain the changeable electrical properties after slaughter as a consequence of shorting through the membrane insulation, here shown as switches shorting capacitors. Although this structural circuit unit is replicated thousands of times between a couple of electrodes a few centimeters apart in a carcass, it allows the impedance of meat to be resolved as a capacitance (C_p) and resistance (R_p) in parallel, as shown in the adjustable components of the Wheatstone bridge (Figure 6.3).

A full analysis of the electrical geometry of this type of model is exceedingly complex, as shown by Gielen et al. (1986), but there are numerous other complexities from biological sources. Intracellular fluid initially has a higher resistance than extracellular fluid (Schanne, 1969), but, as the meat undergoes postmortem changes, there is an efflux

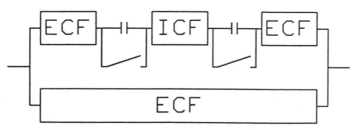

Figure 6.4 *A model for the electrical properties of meat (Swatland, 1980b). ECF and ICF are resistances of extra- and intracellular fluid compartments, respectively, while the muscle fiber membrane is represented by capacitors shorted by switches.*

from the intracellular space into the extracellular space (Heffron and Hegarty, 1974). This change in the resistances is concurrent with changes in membrane capacitance, and we cannot help but measure the two together when we work on-line. Lastly, we must take into account the overwhelming effect of temperature on resistance. Early studies on electrical changes in muscle postmortem (studies that extend back almost a century; Geddes and Baker, 1967) neglected to correct for the inevitable drop and variability in tissue temperature. Temperature is, therefore, of major importance in the on-line evaluation of meat by electrical means. Its effect is primarily on resistance (electrolyte conductance) because cell membranes have only a low temperature coefficient from 0 to 40°C (Fatt, 1964).

Impedance

Just as some of the biology may be unfamiliar to readers coming from an engineering background, meat scientists may be unfamiliar with impedance. For meat, impedance may be visualized as the hypotenuse of a triangle whose base is resistance and whose height is capacitance (capacitive reactance). Capacitance (C_s) and resistance (R_s) resolved as a series circuit may be converted to their parallel equivalents (Geddes and Baker, 1968):

$$R_s = R_p/(1 + (2\pi f C_p R_p)^2)$$

$$C_s = C_p (1 + (1/(2\pi f C_p R_p)^2))$$

Although the intrinsic dielectric anisotropy of meat is very important, there are other practical considerations when making on-line measurements. Laboratory test apparatus may use plate electrodes with an isolated sample between them, but, for the on-line measurement of carcasses, some type of penetration electrode is required. In a simple situation, with just two parallel needle electrodes, there are three ways that the pair may be inserted into the meat (Figure 6.5). These planes make differential contact with the extracellular fluid. In plane b, for example, contact is minimal because muscle fibers have been compressed concentrically, thus increasing the capacitance effect of membranes. In plane c, contact is maximal, and the electrodes may tend to rip open a longitudinal communication channel of extracellular fluid between the electrodes, thus tending to short the test current, which, in turn, reduces the current density on membranes. Plane a tends to be intermediate between the other two planes. Thus, in any system for on-line use, it is important to find the electrode orientation that responds most readily to the subject of measurement and then to standardize it rigorously. With handheld probes, differences between operator usage may well exceed the intrinsic differences between carcasses. Even if a manufacturer fails to communicate this point about the correct use of a probe, the individual user is responsible. There is very little point in taking measurements if they are not controlled as tightly as possible.

Electrode Penetration

Depth of electrode penetration into the meat has a primary effect by altering the area of electrode contact and the current density between the electrodes. With needle electrodes, the contact area may be controlled quite easily with an insulated sleeve covering the base of the electrode,

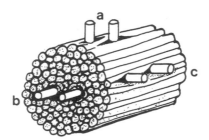

Figure 6.5 *Three ways in which a pair of parallel needle electrodes may be inserted into a meat sample: (a) perpendicular to the fibers and perpendicular to the long axis of the muscle, (b) coaxial with and parallel to both the axes of the muscle fibers and the muscle, and (c) perpendicular to the fibers but along the axis of the muscle.*

so that the exposure to the meat is constant. If a manufacturer does not help the user in this way, then, again, it is essential for the operator to assume responsibility. Likewise, the surface wetness of a carcass, which may vary from dry and sticky to running with spray water, may have a major effect by shorting the electrodes. Carcasses must be damp dry to avoid the penetration electrodes dragging in a variable amount of surface water. But with a completely dry and caked surface, on the other hand, there may be a positive bias on any measurement involving capacitance. Distance between electrodes may be set by a manufacturer, but that does not stop electrodes being bent through careless usage. As would be expected, moving the electrodes together tends to decrease resistance but to increase capacitance (Figure 6.6).

ON-LINE MEASUREMENT OF PORK QUALITY

Pioneer Studies

DFD pork became a problem in England in the 1920s and 1930s when the shipping of pigs by rail to new centralized abattoirs replaced local slaughtering. The penetration of curing ingredients in traditional dry curing is slow and must be done carefully, but when diffusion is hindered even further by a lack of fluid between the muscle fibers in DFD pork, it may lead to spoilage in the center of the ham. Banfield (1935) listed the main factors:

(*1*) Salt concentration in the pickle
(*2*) Extent of pumping and the volume injected into the meat
(*3*) Texture and moisture content of the meat
(*4*) Ambient temperature
(*5*) Available surface area through which the pickle can penetrate (it does not readily penetrate fat)
(*6*) The volumes of meat relative to pickle

He proposed measuring the penetration of curing salts by testing the electrical resistance of the meat (Banfield, 1935; Banfield and Callow, 1935a, 1935b). This led to the discovery that the resistance of fresh pork is quite variable. Perhaps the original thought was simply that, since DFD hams are more dry than normal, they might be detectable by their high electrical resistance.

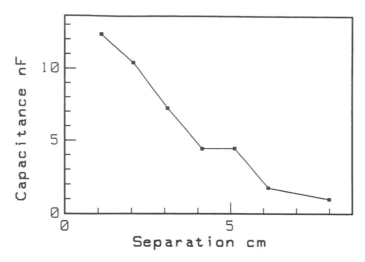

Figure 6.6 *Effect of interelectrode distance on capacitance (Swatland, 1980a).*

Portability

Because of electrode polarization, AC was used, thus measuring impedance, rather than simple resistance. With Banfield's (1935) original apparatus, which used platinized electrodes and a bridge circuit to balance the impedance of the pork against a known impedance (primarily resistive impedance), the null point was found subjectively using earphones to listen to the residual current. But in transferring this technology from the laboratory industry, Banfield (1935) was faced with many of the logistical problems that are with us still today. The key problems are:

(1) Apparatus: (a) weight and size relative to portability, (b) rugged construction, (c) water resistance, (d) power supply, (e) simplicity of operation and data logging, and (f) speed of measurement
(2) Measurement: (a) standardization, (b) sensitivity, (c) accuracy, (d) repeatability (internal error versus sample variation), and (e) degree of objectivity

The solution adopted in Cambridge was to use a handcranked ohmmeter with the output (50 Hz, 92 mA, 80 V) regulated by a slipping clutch mechanism (Callow, 1937). It appears that the equipment was borrowed from soil scientists who used it to measure the water content of soils.

Most of the points listed above for establishing an on-line system are either fairly obvious or are well known to anyone familiar with making scientific measurements. But there are a few hidden subtleties in the list as, for example, in the power supply. The shackle or gambrel that supports the carcass generally is grounded via the overhead rail system: so, too, is one wire of the mains supply. Thus, an off-the-shelf instrument such as a typical impedance meter with a mains power supply cannot be used to probe a hanging carcass without some modification with respect to the grounding of the carcass and (or) the instrument. Furthermore, although abattoirs typically have humid ambient conditions, a hanging carcass soon develops a substantial static charge that complicates the measurement of impedance. Thus, an internal battery power supply has much to offer over mains power and has the added advantage of safety for the operator.

The handcranked ohmmeter was used by Banfield and Callow (1935a, 1935b) in studies to assess factors other than salt concentration that affected impedance. The major factors that are still important in making electrical measurements on meat were discovered:

(1) Pork is electrically anisotropic, with impedance perpendicular to the muscle fibers greater than impedance parallel to the fibers.
(2) Impedance increases with fat content.
(3) Impedance is inversely proportional to temperature.
(4) Impedance changes with the frequency of the test current.

Callow (1938) found that freezing of meat obscured any original differences in impedance, as seems reasonable if ice crystals cut their way through cell membranes, leaving one continuous pool of electrolytes when the muscle is thawed. This provides a simple and convenient method for the detection of meat that has been frozen and thawed, relative to meat that has never been frozen (Salé, 1972).

Relationship with pH

In the 1930s, when DFD pork was a problem for bacon curers because it hindered the already slow penetration of salts in traditional dry-curing methods, it was found that pigs transported to the newly opened central abattoirs produced more DFD than pigs slaughtered on the farm of origin. Also, it was found that impedance was lower in pigs slaughtered on the farm, relative to those at central abattoirs. E. C. Bate Smith, who led the pioneer group of meat researchers at Cambridge, suggested that impedance might be related to pH, and this hypothesis was subsequently tested and confirmed by Callow (1936), as shown in Figure 6.7. Impedance measurements were used to supplement pH measurements and to demonstrate that DFD was caused by transport stress and fighting prior to slaughter. Thus, as far as can be established from the scientific literature, impedance was the first method used for the on-line evaluation of meat, with many advantages over early pH meters. Had it not been for the advent of a world war, the on-line evaluation of pork products might have become a standard industrial practice. As it was, progress in the on-line evaluation of meat, and meat science in general, suffered a setback of many decades.

The early research on the impedance of muscle at high frequencies became quite sophisticated, and comparable results were not to be seen again for forty years. Rowan (1938) and Rowan and Bate Smith (1939) used a Kohlrausch bridge powered by a dynatron oscillator and discovered many subtle aspects of the subject, such as the transient increase in meat impedance soon after slaughter (Figures 6.8 and 6.9). We have not yet proved the basis of this effect, which might be osmotic in origin, even though it is a major source of error in on-line measurements soon after slaughter. Muscle fibers may take up water by osmosis as a result of glycogenolysis, thus reducing the intercellular space (see Plate 13) and increasing the interelectrode resistance. Figure 6.8 shows why on-line measurements soon after slaughter may be unreliable. Some

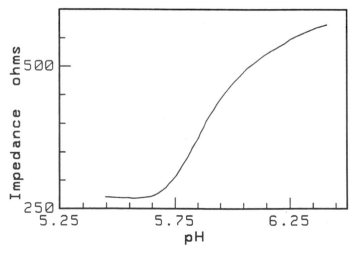

Figure 6.7 A classic discovery: the relationship of electrical impedance at 50 Hz and the ultimate pH of pork (Callow, 1936).

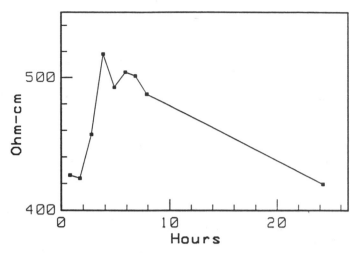

Figure 6.8 *A transient increase in resistance in a quiescent beef sternomandibularis muscle after slaughter. Resistance ultimately declined to a much lower level than shown here (Swatland, 1980a).*

animals do not have an increase in resistance, while, in those that do, it may vary in the time of occurrence. Thus, making measurements at a time when there may or may not be a variable increase in the resistance is likely to lead to disappointing results. Capacitance is more stable than resistance in this regard but still may vary up or down because any change in resistance affects the voltage applied to membrane capacitance. In a parallel circuit, it is difficult to detect capacitance if resistance is very low.

When their studies were interrupted by the war in 1939, Rowan and Bate Smith were only a step away from discovering that capacitance, as well as resistance, is related to pork quality. But their discoveries lay dormant until the late 1970s, when attempts were made to develop more robust methods than glass pH electrodes for sorting and detecting PSE carcasses. There was a flurry of new interest in the subject, much of it appearing to be independent. Work was undertaken in France (Salé, 1972), the Netherlands (van der Wal et al., 1977, 1978), Canada (Swatland, 1980a), and Austria (Pfützner, 1981). The initial intent in the early 1970s was to find ways to identify meat that previously had been frozen or that had been aged or conditioned, while by the end of the decade, interest had switched to the on-line detection of PSE pork. Two on-line electrical methods were developed in 1980 to measure the electrical properties of pork, the MS-Tester and capacitance systems, which are described below. The Carnatest II is a new phase angle impedance system available in Croatia.

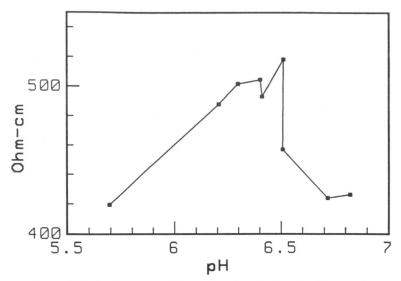

Figure 6.9 *Matching data to Figure 6.8, showing the pH of the muscle.*

Electrical properties also may be used to identify PSE in turkey meat (Aberle et al., 1971) and to quantify the freshness of fish muscle (Kent and Jason, 1975; Jason and Lees, 1971). An instrument called the Torrymeter was developed at the Torry Research Station in Aberdeen, Scotland, as an improvement of a previous instrument called the Intelectron Fish Tester.

MS-Tester

Pfützner (1981) developed a method to measure the dielectric loss factor of isolated pork samples using a frequency synthesizer and an analogue to digital converter, all controlled from a microprocessor. With conductivity γ and dielectric constant ϵ_r, the dielectric loss factor d

$$d \approx \gamma/\epsilon_r$$

was found from the cotangent of the measured phase angle. With isolated samples having a base area of about 20 cm², it was possible to use raster electrodes, two sets of interdigitating rhodium-plated gold lines 1 mm apart etched onto a circuit board. Muscle fibers were parallel to the raster lines of the electrodes.

Pfützner and Fialik (1982) changed the electrode configuration to work with two rhodium-plated electrodes, whose distance apart was regulated by a micrometer screw, each side of disk-shaped samples of gracilis muscle with the fibers parallel to the surface. The optimum frequency for separating subjectively defined PSE and normal pork was found to be 15 kHz.

A portable system called the Testron MS-Tester (MS = meat structure) was developed commercially, using measurements between two parallel scalpel blades 25 mm apart inserted into carcass muscles (Figure 6.10; Pfützner and Fialik, 1982; Kleibel et al., 1983). The intended site of measurement on the hanging carcass was the gluteus medius, which generally is transected in the sirloin when left and right sides of the carcass are separated. However, according to the Testron manual, the blade electrodes also may be used on the longissimus dorsi, provided the blades do not cut transversely across the fibers. Results are displayed on one of nine light-emitting diodes (Table 6.1). This is one solution to a major problem in on-line testing: how to communicate with the human operator, who then must take some sort of action, usually streaming the carcass onto one line or another.

A switch on the rear panel of the handheld instrument (55 × 70 × 180

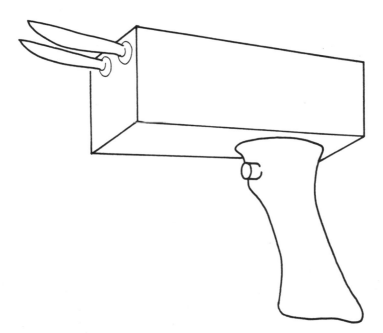

Figure 6.10 *A sketch of the Testron MS-Tester.*

Table 6.1 *Categorization of PSE Pork by Dielectric Loss Factor (d) in the Testron MS-Tester (Kleibel et al., 1983).*

	d	LED Color
A	<2.3	green
B	2.6	green
C	3.1	green
D	4.0	yellow
E	5.6	yellow
F	9.0	yellow
G	15.0	red
H	31.0	red
I	>43.0	red

mm, 980 g) is used to select for measurements at body temperature, room temperature, or meat-cooler temperature. For measurements within three hours of slaughter, green diodes indicate normal pork, yellow for borderline cases, and red for PSE. From three to forty-eight hours after slaughter, normal pork may rate green or yellow, while PSE is again red.

Electrical properties of meat may relate to pH in a nonlinear manner, as shown by the DC voltage output from a modified MS-Tester reported by Seidler et al. (1987) as complex conductivity (Figure 6.11). This was confirmed in the same carcasses ($n = 3028$) with data from another type of meter identified as a LF DIG 550 conductivity meter (Figure 6.12). The inverted nature of the relationships shown in Figures 6.11 and 6.12 originates from the reciprocal nature of the impedance ratios. When capacitance in parallel (C_p) and resistance in parallel (R_p) are related to frequency (f), the quality factor (Q) is defined as

$$Q = 2\pi f C_p R_p$$

The other term in common use is the dissipation factor,

$$D = 1/Q$$

It has been reported that the MS-Tester is capable of differentiating between normal and PSE pork as early as an hour or less postmortem (Pfützner et al., 1981; Kleibel et al., 1983; Chizzolini et al., 1993). However, Schmitten et al. (1983, 1984) were unable to identify strong correlations of the output of the MS-tester at forty minutes postmortem

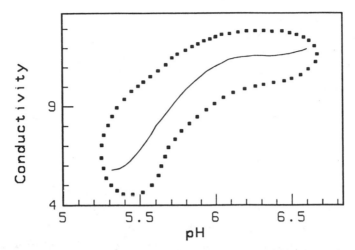

Figure 6.11 *Main axis (line) and extent of scatter (squares) of the relationship of pH₁ with complex conductivity measured by the MS-Tester (Seidler et al., 1987).*

with pH_1, and Schwörer et al. (1984) were unable to demonstrate that the output of the MS-Tester at forty-five minutes postmortem was strongly correlated with meat quality measurements. Fortin and Raymond (1988) evaluated the MS-Tester and the Tecpro QM conductivity meter (from Munich) and were unable to obtain satisfactory predictions of PSE at either sixty minutes or twenty-four hours. Warris

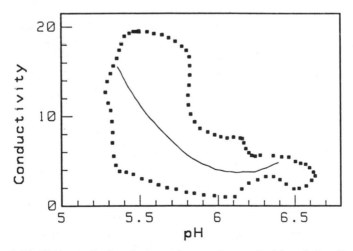

Figure 6.12 *Main axis (line) and extent of scatter (squares) of the relationship of pH₁ with complex conductivity measured by the LF DIG 550 (Seidler et al., 1987).*

et al. (1991) also tested the Tecpro QM conductivity meter, concluding, with a damnation by faint praise, that it was best suited for population quality control, rather than individual carcass grading for PSE. Thus, like other on-line testing for PSE, it would be wise to view measurements made soon after slaughter somewhat critically.

Capacitance

Following the original work by Callow (1936), which was fairly well known in the United States (Moulton and Lewis, 1940) but later forgotten elsewhere, investigations on the electrical properties of pork also were resumed in Canada (Swatland, 1980a, 1980b). The test current frequency (100 kHz) was chosen more out of concern for not activating muscle glycolysis than for finding the most responsive range (Swatland, 1977), and parallel needle electrodes were used so that at least the resistance measurements could be calibrated properly against known conductivity standards. But with capacitance from unknown areas of leaking membranes, no standardization is possible, which is, of course, why Pfützner (1981) chose the dielectric loss factor.

In initial laboratory studies on isolated samples of pork, C_p and R_p were resolved separately with a dual-trace oscilloscope. With constant electrode geometry (parallel needle electrodes), R_p was corrected for temperature and electrode geometry to give resistivity (ohm-cm at $0°C$). Callow's (1937) discovery of a relationship between pH and impedance held true, and both C_p and resistivity were correlated with pH ($r = 0.83$ and $r = 0.71$, respectively). Although sample temperature and electrode orientation had a slight effect on C_p, this was regarded as trivial compared to the effect of temperature on R_p ($r = 0.99$). Thus, to avoid the need to measure meat temperature, C_p was adopted as the most useful element of impedance for the assessment of pork quality. Bearing in mind the practical value of being able to make measurements as soon after slaughter as possible, capacitance also had the highly desirable property of making an early rapid change in the separation between carcasses with normal and fast rates of glycolysis (Figure 6.13).

Battery-operated portable apparatus to measure C_p in the adductor muscles of hanging pork carcasses was developed for use in industry (Swatland, 1981c)—not commercially, but a few instruments for other researchers to use. The apparatus consisted of a meter (14 × 22 × 28 cm, 1 kg) in a chest harness, which enabled measurements to be made

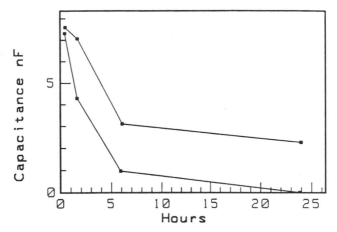

Figure 6.13 *Postmortem decline in capacitance in pork with a normal rate of pH decline (top line, pH >5.8 at 1.5 hours) and in pork with a fast decline (bottom line, pH < 5.8 at 1.5 hours).*

with the needles in one hand and the data to be recorded with the other hand on a writing surface on the top face of the meter. Relationships between capacitance and pH were detected (C_p was correlated with pH_{90} and with pH_u, $r = 0.68$ and $r = 0.58$, respectively; Swatland, 1982a) but, as in the studies of Seidler et al. (1987) on the MS-Tester, it appeared that the relationship of capacitance with pH was nonlinear (Figure 6.14). Of more commercial significance, measurements made on carcasses in the cooler after slaughter could be used to predict the ultimate paleness, as would be seen in hams ($r = -0.88$) and pork chops ($r = -0.84$) several days after slaughter. Also, capacitance in the cooler was correlated with emulsifying capacity (a measure of the value of the pork in processed products) eight days after slaughter, $r = -0.65$.

The apparatus was tested in a commercial environment (Swatland, 1982b). Results were less satisfactory working in a more difficult environment (at line speed) within the range of a commercial population, but correlations of capacitance with pork chop paleness measured objectively (CIE %Y) five days after slaughter were still detectable ($r = -0.37$, $P < 0.01$). Capacitance was correlated with meat exudate at five days, $r = 0.39$ ($P < 0.005$). Independent testing by another research group (Jones et al., 1984) showed that C_p was generally better than pH for the prediction of PSE, with capacitance being correlated with objective paleness, $r = -0.59$.

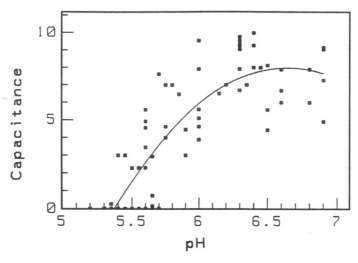

Figure 6.14 *Relationship of capacitance (C$_p$ at 100 kHz) with pH in pork from normal and PSS pigs (from zero to twenty-four hours postmortem).*

Cell Membranes and ATP

Although a direct causal relationship between pH and C_p (such as Figure 6.7) appears to exist and seems quite reasonable scientifically, this relationship withered under critical testing. If lactate-induced damage to ion pumps in the cell membrane is the primary cause of the decrease in membrane capacitance as the pH declines postmortem, then one would expect DFD beef, which has a very high pH not much below that of the living animal, to have very high capacitance. Artificially creating DFD by epinephrine injection, the capacitance of DFD beef was measured: it was lower, not higher, than that of normal beef (Swatland et al., 1982). This prompted a critical examination of the relationships between capacitance, pH, and ATP levels in beef. The correlation of C_p with ATP ($r = .88$) always was consistent, whereas correlations of C_p with pH were weaker and only sporadic in occurrence (Swatland and Dutson, 1984). From this, it was concluded that the postmortem decline of C_p is directly related to a decline in ATP levels (probably by electrical shorting through membranes at the sites of ATP-driven ion pumps), rather than to a direct effect of pH on cell membranes.

The ideal method for real-time measurement of ATP is nuclear magnetic resonance (NMR), so the next experiment was obvious: to examine the relationship between capacitance and ATP levels measured by NMR in pork. As shown in Figure 6.15, capacitance was closely related to ATP. Incidently, the capacitance in Figure 6.15 is high relative to Figures 6.13 and 6.14, because the frequency of the test current was decreased from 100 to 10 kHz to cope with the small size of the core sample required for the small NMR tube. But NMR has advanced considerably since this experiment, and it could now be done on a whole ham. The final step of confirmation was to check the situation on-line in hanging carcasses, where a portable rigorometer was used (Chapter 4). When a muscle finally runs out of ATP postmortem, the myosin molecule heads of the thick myofilaments lock onto the active sites of the actin molecules of the thin filaments. Instead of sliding to produce muscle contraction, the filaments are now locked in rigor mortis. This is detectable in real time by pulling on a strip of muscle to measure its extensibility. As shown in Figure 6.16, as the muscle lost its extensibility, it also lost its capacitance. Both the sliding filament system and the membrane lose their normal physiological function as a consequence of the un-availability of ATP.

The electrical properties of meat certainly are related to pH in some way, as shown in Figures 6.7, 6.9, 6.11, 6.12, and 6.14, but probably only via a mutual or shared relationship with ATP. Dielectric loss factor and capacitance show us when a carcass loses its ATP, and PSE carcasses tend to deplete their ATP more rapidly than normal carcasses, but so may DFD carcasses. In other words, to the best of our present under-

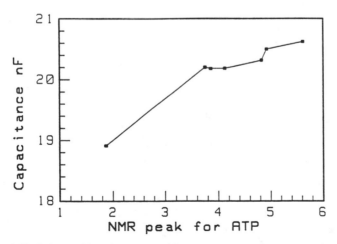

Figure 6.15 *Relationship of capacitance (C_p at 10 kHz) with ATP concentration measured by nuclear magnetic resonance.*

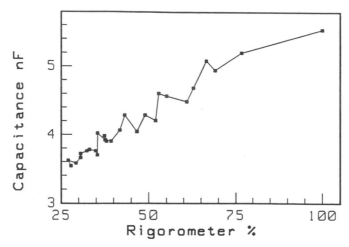

Figure 6.16 *Relationship of capacitance (C_p at 100 kHz) with the progressive loss of adductor muscle extensibility measured with a rigorometer on a hanging pork carcass.*

standing of the situation, appreciable numbers of DFD carcasses could weaken the on-line prediction of PSE using the electrical properties of meat. Fortunately, severe DFD pork carcasses are quite rare. Whether or not the story will change as we learn more about the basic science of this system remains to be seen.

PREDICTION OF LEAN YIELD

TOBEC

Meat is a complex tissue with both resistance and capacitance. When passed at a constant velocity through a magnetic field, it creates a measurable perturbation in the field, although exactly how this occurs is difficult to determine, and the magnetic asymmetry of muscle also may be involved. The principle was used in the EMME (model SA-1, electronic meat measuring equipment, EMME Corp., Phoenix, AZ) for live farm animals, then reappeared as the TOBEC (total body electrical conductivity) HA2 for measuring adiposity in human patients. The manufacturer of the TOBEC HA2 (Agmed Inc., Springfield, IL) then established a subsidiary (Meat Quality Inc.) to produce equipment for on-line evaluation of the fat content of meat. In its latest form, the technology is called the MQI Electromagnetic Scanner. It has taken about twenty years to develop the technology to the point of commercial viability, with much of the technology transfer to the pork industry being accomplished by Forrest and coworkers at Purdue University. Several TOBEC systems installed in the U.S. midwestern pork industry have received high praise from industry and are seen as a tool to stream carcasses for maximum economic return and genetic improvement (NPPC, 1993).

Potential concerns with the technology may be deduced from first principles. It is possible to control or correct for variation in the temperature of the electromagnetic coil, but this is more difficult for meat temperature. The electrical resistance of meat is inversely proportional to temperature, so that a strong temperature effect may be anticipated. There are several factors that may affect the deep temperature of the carcass, such as the extent of exothermic glycolysis, adipose insulation, surface to volume ratio, and refrigeration history, and these may contribute to experimental error in electromagnetic scanning. PSE changes conductivity and might also be a source of error. The shape and position of the sample relative to the field also may affect the response as carcasses or primal cuts are moved through the field, although the serious position effects that led to the rejection of the EMME (Mersmann et al., 1984) appear to have been overcome in the TOBEC. The output of the TOBEC system is difficult to standardize or express in scientific units, and it is very much an interaction between the system and the sample, relative to an interaction of the system with a phantom sample used for standardization. The overall mass of meat passed through the field has a large effect on both the electromagnetic perturbation and the total yield of lean meat, so it is important to ensure that the apparatus is not functioning simply as an elaborate weigh scale, by looking at percent lean and not total lean yield.

On the other hand, the greatest advantage of the technology is that it relates directly to the meat content of the carcass and does not depend on the inverse relationship between subcutaneous fat depth and meat yield, as in optical probe technology. Also important are possibilities for future improvement of the technology, such as scanning at multiple frequencies and using neural networks.

In adapting electromagnetic scanning for the Emscan MQ27 system to measure the fat content of boxed beef on-line in Australia, Sorensen and Treffone (1993) describe their system as having a solenoid coil 1.85 m in length and 0.79 m in diameter, using a frequency of 2.5 MHz. Both the phase angle between the voltage and the current and the amplitude of the current are measured at 50 Hz with a lean meat computation output at 20 Hz.

On-line testing of the MQ25 system is described by Berg et al. (1994), and Figure 6.17 shows a typical electromagnetic scan from their research. As a pork carcass passes through the coil (2.18 m in length and 0.66 m in diameter at 2.5 MHz), a relative energy absorption curve is generated, and the curve is scaled from the height and position of the major peak. In Figure 6.17, the arrows show how the position of the peak (arbitrarily set at 100 on the y-axis) is used to define position 100 on the x-axis and how the 0 position on the x-axis is set at 10% of the peak. This gives a frame of reference to process the otherwise uncalibrated curve, as well as removing noise and effects caused by the initial entry into, and final departure of, the carcass from the field. Areas under parts of the curve and differences between positions on the curve are related to the lean mass in the field (Table 6.2), which enabled Berg et al. (1994) to develop prediction equations for percent carcass lean: $R^2 = 0.830$ in plant A, and $R^2 = 0.863$ in plant B. These results, potentially obtainable at line speeds of 1000 carcasses per hour, were better than those obtainable with an optical probe used under optimal conditions. Essentially, if temperature can be controlled or measured, electromagnetic scanning for pork may be combined with carcass weight

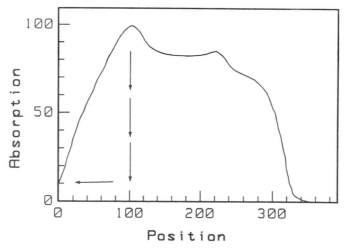

Figure 6.17 *Electromagnetic scan of a pork carcass (Berg et al., 1994). The arrows show how the position of the peak (arbitrarily set at 100 on the y-axis) is used to define position 100 on the x-axis and how the 0 position on the x-axis is set at 10% of the peak.*

Table 6.2 *Simple Correlations of Percent Lean Yield with Areas (A) and Differences (D) Relating to Electromagnetic Scanning Curves for Pork Carcasses Measured at Different Plants at Different Temperatures (Berg et al., 1994).*

Plant A, 2.2 to 9.8°C		Plant B, 41.0 ± 0.01°C	
TOBEC	*r*	TOBEC	*r*
D5—12.5	0.646	D0—17.5	0.750
D7.5—25	0.642	A90—97.5	0.720
D10—27.5	0.670	D97.5—130	0.699
D32.5—50	0.379	D5—32.5	0.734
A130—172.5	0.544	D25—52.5	0.291
A127.5—180	0.543		

to give a strong predictor of total lean ($R^2 = 0.904$, with root mean square error = 1.59 kg; Berg et al., 1994). Gwartney et al. (1994) showed that prediction equations based on electromagnetic scanning of beef sides and quarters may predict 61 to 75 % of the variation in percent lean, although under laboratory conditions, rather than on-line in industry.

Bioelectrical Impedance

Electrode polarization has always been a nuisance in measuring the resistivity of electrolytes and prompted Bouty (1884) to invent the tetrapolar method of measurement, in which two recording electrodes are placed between two source current electrodes. It was developed for medical applications in the 1940s and 1950s, particularly because it is less sensitive to relative movement of the electrodes caused by body movement associated with respiration. Needle electrodes ʻinserted through the skin also bypass the numerous sources of error associated with skin thickness and resistance. A four-terminal bioelectrical impedance analyzer manufactured by RJL Systems (Detroit, MI) has been tested extensively by Marchello, Slanger, and coworkers at North Dakota State University for use in predicting the lean content of animals and meat.

The electrodes (21-gauge needles) may be located in an anterior to posterior sequence along the full length of the animal's back with a 10-cm separation between transmitter and detector electrodes at each end (Slanger et al., 1994). For lamb, prediction equations ($R^2 = 0.97$) for total weight of retail cuts were developed using resistance, reactance, weight, distance between detector electrodes, and temperature. Measured along the side of pork carcasses, the prediction of lean yield ($R^2 = 0.81$) is less accurate than TOBEC but has the advantage that it may be used for both live animals and carcasses (Swantek et al., 1992). Marchello and Slanger (1992) concluded that the method would enable Boston butts to be sorted robotically according to lean content. The method also has been tested satisfactorily on cows for prediction of total muscle from live animals, bled and hot carcasses, and cold carcasses, with R^2 values of 0.90, 0.96, 0.94, and 0.92, respectively (Marchello and Slanger, 1994). Longissimus dorsi intramuscular fat was predicted, $R^2 = 0.83$ (Slanger and Marchello, 1994).

The RJL Systems apparatus, using a current of 800 μA at 50 kHz, has been proven to work satisfactorily, although whether the method can be applied effectively at line speeds on a commercially meaningful narrow range of animals and carcasses remains to be seen. PSE changes conductivity and might be a source of error. Robotic testing of Boston butts looks promising, but whether or not automation of electrode insertion and temperature measurement could be done at a lower cost and with equal or greater reliability than TOBEC remains to be seen. Live animal applications could be advantageous in genetic selection, but the competition is against noninvasive ultrasonic scanning.

REFERENCES

Aberle, E. D. , W. J. Stadelman, G. L. Zachariah, and C. G. Haugh. 1971. *Poultry Sci.,* 50:743.

Arnold, W., R. Steele, and H. Mueller. 1958. *Proc. Natl Acad. Sci.,* 44:1.

Banfield, F. H. 1935. *J. Soc. Chem. Ind. Trans. Comm.,* 54:411T.

Banfield, F. H. and E. H. Callow. 1935a. *J. Soc. Chem. Ind. Trans. Comm.,* 54:413T.

Banfield, F. H. and E. H. Callow. 1935b. *J. Soc. Chem. Ind. Trans. Comm.,* 54:418T.

Berg, E. P., J. C. Forrest, and J. E. Fisher. 1994. *J. Anim. Sci.,* 72:2642.

Bouty, E. 1884. *J. Physique,* 2:325.

Burger, H. C. and R. van Dongen. 1960. *Physics Med. Biol.,* 5:431.

Callow, E. H. 1936. *Special Rep. Dept. Sci. Indust. Res., Food Invest. Board, London,* 75:75.

Callow, E. H. 1937. *Ann. Rep. Food Invest. Board, London,* pp. 46–49.

Callow, E. H. 1938. *Ann. Rep. Food Invest. Board, London,* pp. 55–59.

Catton, W. T. 1957. *Physical Methods in Physiology.* London: Sir Isaac Pitman and Sons.

Chizzolini, R., E. Novelli, A. Badiani, P. Rosa, and G. Delbono. 1993. *Meat Sci.,* 34:49.

Cole, K. S. 1970. Dielectric properties of living membranes, in *Physical Principles of*

Biological Membranes, F. Snell, J. Wolken, G. Iverson, and J. Lam, eds., New York: Gordon and Breach Science Publishers, pp. 1–15.

Conway, E. J., H. Geoghegan, and J. I. McCormack. 1955. *J. Physiol.,* 130:427.

Echeverria, O. M., G. H. Vazquez-Nin, and J. G. Ninomiya. 1983. *Acta Anat.,* 116:358.

Epstein, B. R. and K. R. Foster. 1983. *Med. Biol. Eng. Comput.,* 21:51.

Faraday, M. 1845. *Experimental Researches in Electricity, Vol. III,* Series XX, pp. 27–53. Reprinted by Dover Publications, New York, 1965.

Fatt, P. 1964. *Proc. Roy. Soc., Lond.,* B 159:606.

Fortin, A. and D. P. Raymond. 1988. *Can. Inst. Food Sci. Technol. J.,* 21:260.

Geddes, L. A. and L. E. Baker. 1967. *Med. Biol. Engng.,* 5:271.

Geddes, L. A. and L. E. Baker. 1968. *Principles of Applied Biomedical Instrumentation.* New York: John Wiley & Sons.

Gielen, F. L. H., H. E. P. Cruts, B. A. Albers, K. L. Boon, W. Wallinga de Jonge, and H. B. K. Boom. 1986. *Med. Biol. Engin. Comput.,* 24:34.

Gill, C. O., N. G. Leet, and N. Penney. 1984. *Meat Sci.,* 10:265.

Guttman, R. 1939. *J. Gen. Physiol.,* 22:567.

Gwartney, B. L., C. R. Calkins, R. S. Lin, J. C. Forrest, A. M. Parkhurst, and R. P. Lemenager. 1994. *J. Anim. Sci.,* 72:2836.

Hamdy, M. K., L. E. Kunkle, and F. E. Deatherage. 1957. *J. Anim. Sci.,* 16:490.

Heffron, J. J. A. and P. V. J. Hegarty. 1974. *Comp. Biochem. Physiol.,* 49A:43.

Jason, A. C. and A. Lees. 1971. Estimation of fish freshness by dielectric measurement, Report T71/7. Aberdeen, Scotland: Torrey Research Station.

Jennische, E., E. Enger, A. Medegard, L. Appelgren, and H. Haljamae. 1978. *Circ. Shock.,* 5:251.

Jones, S. D. M., A. Fortin, and M. Atin. 1984. *Can. Inst. Food Sci. Technol. J.,* 17:143.

Kent, M. and A. C. Jason. 1975. Dielectric properties of foods in relation to interactions between water and the substrate, in *Water Relations of Food,* R. B. Duckworth, ed., New York: Academic Press, pp. 221–232.

Kleibel, A., H. Pfützner, and E. Krause. 1983. *Fleischwirtshaft,* 63:1183.

Marchello, M. J. and W. D. Slanger. 1994. *J. Anim. Sci.,* 72:3118.

Mersmann, H. J., L. J. Brown, E. Y. Chai, and T. J. Fogg. 1984. *J. Anim. Sci.,* 58:85.

Moulton, C. R. and W. L. Lewis. 1940. *Meat through the Microscope.* University of Chicago: Institute of Meat Packing.

NPPC. 1993. Scanning for gold, in *Porkreport,* 12(2):14. Des Moines, IA: National Pork Producers Council.

Pfützner, H. 1981. *Vth International Conference on Electrical Bioimpedance.* Tokyo, paper 56.

Pfützner, H., and E. Fialik. 1982. *Zbl. Vet. Med. A.,* 29:637.

Pfützner, H., E. Fialik, E. Krause, A. Kleibel, and W. Hopferwieser. 1981. *Proc. 27th Eur. Meeting Meat Res. Workers,* Vienna, p. 50.

Rowan, A. N. 1938. *Ann. Rep. Food Invest. Board, Lond.,* p. 23.

Rowan, A. N. and E. C. Bate Smith. 1939. *Ann. Rep. Food. Invest. Board, Lond.,* p. 11.

Salé, P. 1972. Appareil de détection des viandes décongelées par mesure de conductance électrique. *Suppl. Bull. Inst. Internat. Froid.* 2:265–275.

Schanne, O. F. 1969. Measurement of cytoplasmic resistivity by means of the glass microelectrode, in *Glass Microelectrodes,* M. Lavallee, O. F. Scanne, and N. C. Hebert, eds., New York: John Wiley & Sons.

Schmidt, G. R., G. Goldspink, T. Roberts, L. L. Kastenschmidt, R. G. Cassens, and E. J. Briskey. 1972. *J. Anim. Sci.*, 34:379.

Schmitten, F., K. H. Schepers, A. Festerling, B. Hubbers, and U. Reul. 1983. *Schweinzucht Schweinemast.*, 12:418.

Schmitten, F., K. H. Schepers, H. Jüngst, U. Reul, and A. Festerling. 1984. *Fleischwirtshaft*, 64:1238.

Schwan, H. P. and C. F. Kay. 1957. *Ann. N. Y. Acad. Sci.*, 65:1007.

Schwörer, D. and J. K. Blum. 1983. *Kleinviehzuchter*, 31:278.

Schwörer, D., J. K. Blum and A. Rebsamen. 1980. *Livest. Prod. Sci.*, 7:337.

Schwörer, D., J. K. Blum, and A. Rebsamen. 1984. In *Biophysical PSE-Muscle Analysis*, H. Pfützner, ed., Austria: Technical University of Vienna, D-2, p. 284.

Seidler, D., E. Bartnick, and B. Nowak. 1987. In *Evaluation and Control of Meat Quality*, P. V. Tarrant, G. Eikelenboom, and G. Monin, eds., Dordrecht, the Netherlands: Martinus Nijhoff Publishers, pp. 175–190.

Slanger, W. D. and M. J. Marchello. 1994. *J. Anim. Sci.*, 72:3124.

Slanger, W. D., M. J. Marchello, J. R. Busboom, H. H. Meyer, L. A. Mitchell, W. F. Hendrix, R. R. Mills, and W. D. Warnock. 1994. *J. Anim. Sci.*, 72:1467.

Sorensen, B. and G. Treffone. 1993. *Food Australia*, 45(7):33.

Swantek, P. M., J. D. Crenshaw, M. J. Marchello, and H. C. Lukaski. 1992. *J. Anim. Sci.*, 70:169.

Swatland, H. J. 1977. *Can. Inst. Food Sci. Technol. J.*, 10:280.

Swatland, H. J. 1980a. *J. Anim. Sci.*, 50:67.

Swatland, H. J. 1980b. *J. Anim. Sci.*, 51:1108.

Swatland, H. J. 1980c. *Can. Inst. Food Sci. Technol. J.*, 13:45.

Swatland, H. J. 1981a. *Can. Inst. Food Sci. Technol. J.*, 14:147.

Swatland, H. J. 1981b. *Meat Sci.*, 5:451.

Swatland, H. J. 1981c. *J. Anim. Sci.*, 53:666.

Swatland, H. J. 1982a. *Can. Inst. Food Sci. Technol. J.*, 15:92.

Swatland, H. J. 1982b. *Can. J. Anim. Sci.*, 62:725.

Swatland, H. J. 1985. *J. Anim. Sci.*, 61:887.

Swatland, H. J. 1986. *Comput. Electron. Agric.*, 1:271.

Swatland, H. J. 1994. *Structure and Development of Meat Animals and Poultry.* Lancaster PA: Technomic Publishing Co., Inc.

Swatland, H. J. and T. R. Dutson. 1984. *Can. J. Anim. Sci.*, 64:45.

Swatland, H. J., P. Ward, and P. V. Tarrant. 1982. *J. Food Sci.*, 47:686.

van der Wal, P. G., H. L. van Boxtel, A. van Velthuysen, and T. P. Dekker. 1977. *Elektrische Geleidbaarheid in Vlees. Een nadere oriëntering.* Instituut voor Veeteeltkundig Onderzoek 'Schoonoord', Zeist. Rapport C-312.

van der Wal, P. G., T. P. Dekker, H. L. van Boxtel, and G. van Essen. 1978. Electrical conductivity; A meat quality criterion? *European Meeting Meat Research Workers*, Kulmbach, Germany.

Warris, P. D., S. N. Brown, and S. J. M. Adams. 1991. *Meat Sci.*, 30:147.

Winger, R. J. and C. G. Pope. 1981. *Meat Sci.*, 5:355.

Yawo, H. and M. Kuno. 1983. *Science*, 222:1351.

CHAPTER 7

Meat Color

INTRODUCTION

MEAT COLOR IS extremely important commercially because most customers have strongly developed color preferences. The average person prefers bright colored meat, while connoisseurs search out the browns and dull colors of meat that has been aged to the peak of gastronomic perfection. In Chapter 5, we considered the optical properties of meat, the basic microstructural features of meat responsible for light scattering, but without considering the pigments that give meat its color. Chromophores create the subjective sensation of meat color in a system that is otherwise a translucent, optically anisotropic tissue. Thus, pH-dependent light scattering in meat has a strong effect on the overall paleness or darkness of meat, regardless of the concentration or state of chromophores. In meat washed to remove chromophores, reflectance is almost a linear function of wavelength (from about 0.3 at 420 nm to about 0.75 at 700 nm). This chapter explains how myoglobin and its derivatives are primarily responsible for selective absorbance in meat.

INSTRUMENTATION

Reflectance attachments for cuvette spectrophotometers and integrating-sphere colorimeters are, or were, in common use for the study of meat color (Francis and Clydesdale, 1975), and these were the typical instruments that were used for meat research in the laboratory. Except for a few transient applications in the area of meat packaging, there was no hope of adapting such instruments for on-line use to measure meat and fat color. With the advent of fiber optics and photodiode-array technology, however, a new generation of commercial reflectance spectrophotometers has become available, many of which may be used on-line.

185

The basic choice when acquiring or developing new on-line instrumentation is whether to use a self-contained portable instrument in which all the optics are peripheral and the connection to the central processor is via a conventional computer linkage, such as an RS-232 port using ASCII character codes (Swatland, 1986b). The primary advantages are portability for use on intact carcasses and decentralization, which allows repairs, flexibility, and expansion of the system. The alternative is to use optical fibers to carry optical information from the peripheral application to a central optical processor. This allows the peripheral units to be relatively simple and rugged, because all the precision optics are safely installed in the central processor, and vendors of such systems have succeeded in making complex multivariate analysis programs for spectra readily available to the user (Schirmer and Gargus, 1986).

With the advent of fiber-optic systems, there are several new ways in which the microstructure of meat may be optically coupled to spectrophotometers. To the conventional format of an air-filled interface between the meat surface and the optical system are other formats such as direct contact between muscle fibers and optical fibers, first introduced by MacDougall (1980) for monochromatic reflectance measurements (Chapter 3). Direct contact with diodes also may be used in some commercial monochromatic systems for measuring meat paleness. The protocols for the precise measurement of color specify how the sample should be illuminated and how the resulting reflectance should be measured. But taking similar optical components (illuminator and photometer) and placing them in direct contact with the sample, without an airspace between them, changes the results enormously. As yet, there has been no general agreement regarding how to cope with this problem.

If the incident illumination contains a high intensity of UV light around 365 nm, adipose and connective tissues of meat fluoresce with emissions from 400 to around 520 nm. Under most situations, however, the bias imposed on reflectance spectra by fluorescence is imperceptible. For uniformity, the data presented here are shown directly in reflectance units (on a variable scale), rather than in a logarithmic transformation (logarithm to the base 10 of the reciprocal of reflectance, R_λ) as used by some authors.

Unfortunately, in many published reports, the details of experimental conditions are lacking. A number of factors affect meat color, such as:

(1) Exposure of animals or their meat to nitrite or carbon monoxide
(2) Animal age, breed, sex, and nutritional status

(*3*) Animal stress during transport and slaughter

(*4*) Carcass refrigeration rate and postmortem aging

(*5*) Variation in atmospheric exposure, ambient temperature, and relative humidity at the time of measurement

Nitrite and carbon monoxide are particularly troublesome. Nitrite may originate from contact with cured meat juice or spices or from the water added in processing; while carbon monoxide may be acquired by the live animal from vehicle exhaust fumes or by meat being exposed to oven gases. Meat is also such a variable commodity with respect to muscle fiber arrangement, pH, and connective tissue content that it is difficult to standardize the way in which measurements are made. Reflectance spectra from intact meat should not, therefore, be judged with the same rigorous standards that might be used in biochemical spectroscopy. Although standardization for the reflectance spectrophotometry of intact meat may be less precise than in other fields, the subject is of vital practical importance. Reflectance spectra measured rapidly and nondestructively on intact carcasses may provide a realistic method for the on-line measurement of meat color, and color has a tremendous commercial importance.

Conventional reflectance spectrophotometers and colorimeters may be considered together since they usually consist of a light source incident upon the sample through an air-filled space above the sample surface. The incident light and the photometer usually are arranged at an angle to exclude specular reflectance as far as possible. Light is scattered by the meat microstructure and selective absorbance occurs in the subsurface volume of the sample. The area of the sample that is measured generally ranges from about 0.5 cm² with reflectance attachments for spectrophotometers to several square centimeters for colorimeters with spectrophotometric capabilities. Thus, to distinguish this type of apparatus from optical fiber or microscope systems that measure only very small amounts of meat, the term *macroscopic* is used.

The method of illumination and the relatively large area measured both facilitate standardization. Thus, a newly pressed plate of barium sulfate powder calibrated against magnesium oxide by the supplier may be used to prepare a secondary standard that, in turn, may be used to calibrate a robust working standard such as a plate of vitrolite, enamel, or opal glass.

MYOGLOBIN

By 1900, it was apparent that the red pigmentation of skeletal muscles in many vertebrates originates from myoglobin, rather than hemoglobin, although the name *myoglobin* was not adopted until the 1920s (Kagen, 1973). Myoglobin is highly soluble and is formed from a single polypeptide chain twisted around an oxygen-carrying heme group composed of an atom of iron and a porphyrin ring. The oxygen storage capacity of myoglobin is quite limited in farm animals, and the primary function of myoglobin is to transport oxygen within the muscle fiber. Myoglobin concentration in skeletal muscle does not change appreciably with physiological adaptation to high altitude, exercise training, or level of nutrition.

During exsanguination, circulating hormones may cause vasoconstriction (Hall et al., 1976) so that meat from animals exsanguinated in an abattoir sometimes contains very small amounts of hemoglobin from residual erythrocytes (Warriss, 1977). However, the concentration of myoglobin is by far the most important factor in determining the redness of meat. Differences in the redness of meat from different species and parts of the carcass are caused by differences in myoglobin concentration in the live animal (Lawrie, 1985). Differences in myoglobin distribution also may be quite distinct at the cellular level. In pork, for example, myoglobin is confined to the aerobic muscle fibers that occur axially within muscle fasciculi (Morita et al., 1969), as shown in Plate 10. Thus, postural muscles of a pork carcass may appear darker than those used for rapid locomotion because the postural muscles contain more fibers with a high myoglobin concentration.

Muscle fibers with a high myoglobin concentration may range from completely aerobic to a dual aerobic-anaerobic capability in their histochemical profiles. Pale muscles in a pork, veal, or poultry carcass usually contain a high proportion of anaerobic fibers with high levels of stored glycogen (unless depleted by struggling) and appropriate pathways for rapid glycolysis. Fibers with a slow speed of contraction are predominantly aerobic. Thus, muscle color may be related to ATPase activity, so that a dark red color generally indicates slow ATPase activity, and, in pork, color differences within muscles may be related to differences in muscle metabolism postmortem (Beecher et al., 1965). Differences in myoglobin concentration between fiber types are less well developed in species with uniformly dark meat, and problems with variegated cuts of meat are only conspicuous in pork.

Metmyoglobin formation is very important in meat packaging and display, but there are many subtleties to the browning of meat that is always a concern at the meat counter. Cytochromes affect meat color indirectly, because the rate at which they utilize oxygen determines the depth to which oxygen is found in the meat. In the absence of oxygen, cytochromes may reduce metmyoglobin back to myoglobin. Thus, the effects of histochemical fiber types on meat color are more complicated than the simple relationship between myoglobin distribution and histochemical fiber types. Peroxisome (microbodies) with catalase and peroxidase activity also occur in muscle fibers (Hand, 1974), and red muscles may have higher catalase activity then white muscles (Jenkins and Tengi, 1981). Glutathione peroxidase influences metmyoglobin formation (Lin and Hultin, 1978).

The major objective of early research on meat spectrophotometry during the 1950s was to develop methods to quantify the relative amounts of deoxymyoglobin, oxymyoglobin, and metmyoglobin. This was to facilitate the packaging research that was undertaken as the retail meat industry adopted refrigerated supermarket display counters for prepackaged meat cuts (Bailey, 1968). Macroscopic reflectance spectrophotometers, as described above, were used to establish the general relationship between pigment chemistry and the reflectance spectra of meat.

Deoxymyoglobin and Metmyoglobin

The transformation of purple deoxymyoglobin to bright red oxymyoglobin may be seen a few minutes after cutting into the anaerobic center of a piece of beef, but it is much slower in pork and chicken (Millar et al., 1994). After prolonged exposure to the atmosphere, however, the iron atom of myoglobin may be converted to the ferric form, and brown metmyoglobin is formed. This occurs most rapidly just under the surface of the meat. Figure 7.1, taken from Hunt (1980), is similar to the widely accepted reflectance spectra of myoglobin derivatives found in textbooks (Francis and Clydesdale, 1975; Schwimmer, 1981). However, it is not altogether clear how these fundamentally important reflectance spectra were obtained in the reflectance mode from purified myoglobin derivatives. The absorbance spectra of Francis and Clydesdale (1975) are shown as extinction coefficients, and those of Hunt (1980) are shown as R_A values (without a bibliographic source or an indication of the instrumentation used). Schwimmer (1981) used the reflectance spectra given by Strange et al. (1974), but the curvilinear spectra of Strange et al. (1974) extend smoothly from 400 to 700 nm whereas the spectrophotometer used by these authors was a Beckman DU model 2400 with a minimum wavelength of 520 nm and only nine other wavelengths. The spectra presented by Snyder (1968) start just below 450 nm, while those of another possible source (Broumand et al., 1958) stop at 460 nm. Thus, as far as may be determined from these sources, the reflectance spectra for deoxymyoglobin, oxymyoglobin, and metmyoglobin that are fundamental to our understanding of meat spectrophotometry might, perhaps, not all have originated by direct measurement, but by the transformation of the absorbance spectra of purified pigments in solution.

Even the original sources of the myoglobin absorbance spectra used by meat scientists are difficult to trace. The spectra published by Bowen (1949) may have been passed on by Broumand et al. (1958); however, this does not explain the source of the oxymyoglobin spectrum given by Broumand et al. (1958), which is attributed to Bowen (1949) yet not shown by the latter author. Other possible original sources are Kiese and Kaeske (1942) or Theorell and de Duve (1947). Reflectance spectrophotometry of intact skeletal muscle appears to have been pioneered by Ray and Paff (1930) in an attempt to avoid chemical alteration of myoglobin during its prolonged extraction with caustic reagents, but the valuable information so obtained was apparently overlooked by meat scientists.

Although the spectra in Figure 7.1 have been adjusted (by an earlier author) to have an isobestic point at 525 nm, the general features are similar to those that may be observed in a sample of meat as deoxymyoglobin becomes oxygenated and then oxidized. The Soret absorbance bands of deoxymyoglobin, oxymyoglobin, and metmyoglobin are at 434, 416, and 410 nm, respectively (Morton, 1975; Bertelsen and Skibsted, 1987) but are not clearly visible in Figure 7.1, even in the original logarithmic transformation (R_A).

The oxygenation of deoxymyoglobin causes a characteristic change that involves the loss of the absorbance band at 555 nm and the appearance of two new absorbance bands at 542 and 578 nm (Fleming et al., 1960; Morton, 1975; Bertelsen and Skibsted, 1987) with millimolar absorptivities of 13.2 and 13.3, respectively (Morton, 1975). For bovine oxymyoglobin in phosphate buffer, Wolfe et al. (1978) found that oxymyoglobin millimolar absorptivities at 542 and 580 nm were both 14.4. In older literature, these two bands are reported at slightly higher

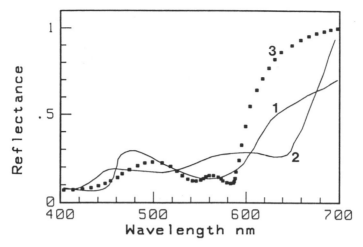

Figure 7.1 *Reflectance spectra of myoglobin (1), metmyoglobin (2), and oxymyoglobin (3) [from Hunt (1980)].*

wavelengths, for example, 581 nm for the alpha band (Millikan, 1939). By reflectance spectrophotometry, Ray and Paff (1930) found that the 542- and 578-nm absorbance bands were of equal intensity. The magnitudes of these two oxymyoglobin bands provide a way to trace how published spectra have been obtained. In meat that has not been exposed to a rapid or extensive decline in pH postmortem or treated with chemical reagents to stabilize pigment chemistry, these two bands generally are of equal intensity when measured by reflectance from intact meat. When measured by absorbance spectrophotometry of extracted pigments, absorbance at 542 nm usually is less than at 578 nm. A likely explanation is that intact meat containing oxymyoglobin also probably contains metmyoglobin. The oxidation of oxymyoglobin to metmyoglobin is accompanied by a loss of absorbance at 578 nm relative to absorbance at 542 nm (Krzywicki, 1979; Bertelsen and Skibsted, 1987). Hence, a natural balance of oxymyoglobin and metmyoglobin generally results in equal absorbance bands when measured by reflectance from intact meat; however, this is not totally reliable. Snyder (1965), for example, measured the reflectance of chemically stabilized myoglobin derivatives on intact samples of beef using a Beckman DK2A so that, if borrowed by later authors, these spectra would have had the appearance of being transformed from absorbance spectra on purified pigment.

Oxidation of myoglobin to metmyoglobin causes other profound changes in the reflectance spectrum. From absorbance spectrophotometry of purified metmyoglobin (Bertelsen and Skibsted, 1987), it appears possible that the secondary reflectance peak of metmyoglobin around 600 nm might include an additional small peak at 565 nm (not seen in Figure 7.1). A general feature to note in Figure 7.1 is that, apart from the curvilinear features just described, reflectance tends to be proportional to wavelength so that the 400-/700-nm ratio of reflectance is low.

Absorbance versus Scattering

Following Broumand et al. (1958), considerable effort was directed towards the estimation of deoxymyoglobin, oxymyoglobin, and metmyoglobin ratios from a small number of monochromatic reflectance measurements made on intact meat. Attempts to deal with the scattering properties of meat were made by adopting the Kubelka-Munk function (Dean and Ball, 1960) but were based on the untested assumption that the scattering coefficient (S) of meat is constant and does not need to be measured on thin slices of meat after the reflectance has been measured (Stewart et al., 1965). In this notation, K is the absorbance coefficient. Snyder (1968) argued (correctly) that it makes no sense to use K/S ratios to plot whole spectra without taking into account the fact that scattering may vary with wavelength, as in the work of Satterlee and Hansmeyer (1974). Thus, at each wavelength measured, K and S should be determined separately and not from the Kubelka-Munk function:

$$K S^{-1} = (1 - R)^2 (2R)^{-1}$$

The appropriate wavelengths for estimating the proportions of myoglobin derivatives are given by Hunt (1980). The critical comments of Snyder (1968) are recommended to anyone who might use them, particularly concerning the special requirements for the reflectance standard that is required (Snyder and Armstrong, 1967).

REFLECTANCE SPECTRA

Beef

Faustman and Cassens (1991) found that the rate of metmyoglobin accumulation in beef varies between breeds, as well as between types of muscle. Reflectance spectra of beef are shown in Figure 7.2. Data

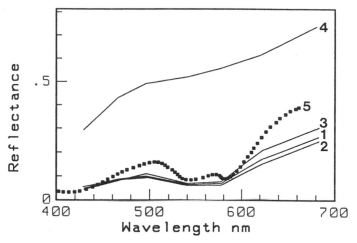

Figure 7.2 *Macroscopic reflectance spectra of beef longissimus dorsi at zero, six, and twenty-four hours postmortem (spectra 1, 2, and 3, respectively) after removal of water-soluble chromophores (4), and measured on tenderloin steaks (5). Data were collected with a Zeiss Elrepho [spectra 1 to 4 from Swatland (1982)] and with the reflectance attachment of a Beckman DR [spectrum 5 from Tappel (1957)].*

collected with a Zeiss Elrepho colorimeter (which is based on an integrating sphere) were close to those collected with a Beckman DR spectrophotometer with a reflectance attachment, particularly with regard to the magnitude of reflectance units. The formation of oxymyoglobin is apparent in spectrum 5 of Figure 7.2 collected with the Beckman but would have been beyond the resolution of the discrete filter monochromator of the Elrepho. Although the Elrepho is very limited as a spectrophotometer, its powder-press accessories for preparing barium sulfate standards calibrated against magnesium oxide make it particularly reliable for standardized reflectance measurements. Spectrum 4 in Figure 7.2 shows the reflectance of beef after extensive washing to remove water-soluble chromophores and gives some idea of the magnitude of reflectance by scattering from structural proteins without subsequent selective absorbance by dissolved chromophores, although, even with drastic treatment such as homogenization, it is difficult to remove completely all trace of the Soret band (Solberg, 1970). Although myoglobin is readily soluble, organelles containing hemoproteins, such as mitochondria, are more difficult to remove.

From the fairly close agreement of the Elrepho and Beckman DR, the range between spectra 2 and 5 in Figure 7.2 gives a reasonable idea of the reflectance spectra of typical beef relative to magnesium oxide.

Pork

The muscles of pigs reared under constant temperature conditions contain more myoglobin than those of pigs reared in fluctuating temperature conditions (Thomas and Judge, 1970), and heritability estimates for pork muscle color range from 0.17 to 0.55 (Pease and Smith, 1965). At the same site within certain muscles, there may be differences in muscle color between breeds of pigs (Kastenschmidt et al., 1968).

Macroscopic reflectance spectra of pork are shown in Figure 7.3. The range for normal pork with relatively low (gluteus medius) and high (iliopsoas) myoglobin concentration calibrated against magnesium oxide with an Elrepho is shown by spectra 1 and 2, respectively, in Figure 7.3 (Swatland, 1982). Within a carcass, the differences between spectra from muscles that differ in myoglobin concentration may increase with animal age and may be far more conspicuous than shown in Figure 7.3. The other spectra in Figure 7.3 were all collected with a Spectro-Computer. Differences in reflectance between PSE and DFD are shown in Figure 7.3 by spectra 3 and 6 for PSE, spectra 4 and 7 for normal pork, and spectra 5 and 8 for DFD. Spectra 3 to 5 were collected with four

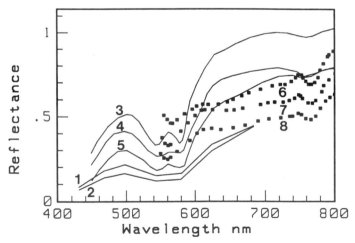

Figure 7.3 Macroscopic reflectance spectra of pork. Spectra 1 and 2 are for gluteus medius and iliopsoas, respectively, measured with a Zeiss Elrepho [from Swatland (1982)]. Spectra 3 to 8 were all obtained with a Spectro-Computer. Spectra 3, 4, and 5 are for PSE, normal, and DFD longissimus dorsi muscles, respectively [from Davis et al. (1978)]. Spectra 6, 7, and 8 are for PSE, normal, and DFD longissimus dorsi muscles, respectively [from Birth et al. (1978)].

equidistant silicon detectors arranged around the incident light path, which was perpendicular to the sample surface. The reflectance standard was vitrolite with a reflectance of 93% relative to barium sulfate (Davis et al., 1978). Unfortunately, the reflectance of vitrolite usually decreases towards 700 nm. This bias may have contributed a little to the high reflectance values of spectra 3 to 5 in Figure 7.3, reported by Davis et al. (1978).

The Spectro-Computer reflectance spectra for pork reported by Birth et al. (1978) are somewhat less intense at many wavelengths (spectra 6 to 8 in Figure 7.3) but are not really much closer to the Elrepho spectra (spectra 1 and 2 in Figure 7.3). Birth et al. (1978) used an integrating sphere and a barium sulfate standard. Sample variation in myoglobin concentration and pH may have contributed to the separation between these two sets of spectra (Elrepho, 1 and 2; and Spectro-Computer, 6 to 8).

The considerable divergence between the three sets of spectra for pork shown in Figure 7.3 makes it difficult to suggest where the reflectance spectra for typical pork might be located. Anywhere between spectra 2 and 7 might be reasonable. This is confirmed by the work of Elliott (1967), as shown in Figure 7.4. The optical anisotropy of pork that is evident in Figure 7.4 was discussed earlier in Chapter 5.

Fiber-Optic Reflectance

A bundle of optical fibers (light guide) composed of a common trunk bifurcating to two branches may be used to obtain reflectance spectra from meat. The importance of this method is that measurements may be made on-line from carcasses without slicing the meat to produce a flat surface (as is required for macroscopic systems). By measuring internal reflectance (microstructural light scattering minus selective absorbance by chromophores), problems with surface discoloration are avoided. Specular reflectance from the wet surface of cut meat is also excluded.

The common trunk of the bifurcated light guide connects directly to the meat sample, while the light source connects to one branch and the photometer connects to the other branch (Figures 3.10 and 3.11). The effects of any unavoidable ambient illumination may be minimized by locating the monochromator on the outgoing branch of the light guide. Since reflected light is collected by the distal ends of the optical fibers (a two-dimensional area), the real parameter of measurement may be sterance ($Wsr^{-1} m^{-2}$) rather than exitance (Wm^{-2}), as outlined by Mims (1975). This creates a problem in terminology. Those used to working

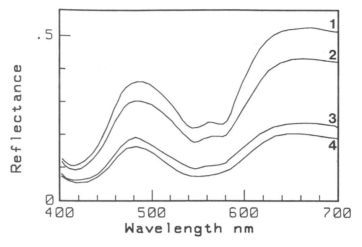

Figure 7.4 Macroscopic reflectance spectra of pork [from Elliott (1967)]. Spectra 1 and 2 are for pale pork, and spectra 3 and 4 are for dark pork. Spectra 1 and 3 are with muscle fibers parallel to the sample surface, and spectra 2 and 4 are with fibers perpendicular.

with conventional reflectance spectrophotometry may object, quite correctly, to the light gathered by an optical fiber directly from a tissue being called reflectance. The term *sterance* might be used, but even this does not take into account the fact that the wavelengths that exit and enter optical fibers are affected by the refractive indices at the end of the fiber. It has been proposed that a new term, *interactance*, should be used to denote the light that is returned to an optical fiber after being scattered or reflected by a tissue (Conway et al., 1984), but this suggestion does not appear to have been widely accepted. Thus, in this book, when reflectance is measured through optical fibers, it is called fiber-optic reflectance.

Many potential problems (such as selective absorbance by optical fibers) may be avoided by a judicious standardization protocol. The main problem that has not yet been solved, however, is how to standardize the system to 100% reflectance at each wavelength in a manner that simulates the optical conditions that occur when the optical fibers are connected directly to the meat sample. Ideally, the refractive indices of the media at the distal ends of the optical fibers should be constant during both standardization and measurement (Chapter 3).

A comparison of macroscopic reflectance spectra with fiber-optic reflectance spectra on the same samples is shown in Figure 7.5 where, for reference, the dotted lines show the spectra carried forward from previous figures (spectra 4, 5, and 6 are Elrepho spectra of washed beef, pork iliopsoas, and beef at twenty-four hours postmortem, respectively). For washed beef lacking any easily removable chromophores in the microstructure, the fiber-optic reflectance spectrum (spectrum 1 in Figure 7.5) is very similar to the Elrepho reflectance spectrum.

For intact pork and beef (spectra 2 and 3, respectively, in Figure 7.5), the fiber-optic reflectance spectrum is less intense than the Elrepho reflectance spectrum, except towards 700 nm. It would appear, therefore, that the mean path length is increased for fiber-optic reflectance spectra so that, when chromophores are present, greater selective absorbance occurs. Since the primary chromophore, myoglobin (Figure 7.1), has very low absorbance towards 700 nm, macroscopic reflectance and fiber-optic reflectance spectra tend to converge towards 700 nm.

MYOGLOBIN CONCENTRATION

Myoglobin concentration is an important attribute of various types in meat, as in the two-toning effect sometimes seen in various ham muscles

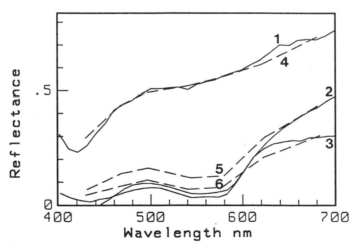

Figure 7.5 *Fiber-optic reflectance spectra. Fiber-optic spectra 1, 2, and 3 are for washed beef, pork, and beef, respectively [from Swatland (1982)]. For reference, macroscopic reflectance spectra 4, 5, and 6 for washed beef, pork, and beef are carried forward from spectrum 4 of Figure 7.2, spectrum 2 of Figure 7.3, and spectrum 3 of Figure 7.2, respectively.*

and in the conspicuous color differences between chicken breast and leg meat. The redder meats carry with them a variety of associated properties, such as higher lipid content and a greater propensity for rancidity, so that monitoring meat color is an advantage in an automated meat-processing operation.

Veal Color

Veal coloration sometimes is contentious because of premiums paid for pale veal. Dietary iron deficiency reduces the activity of a number of respiratory enzymes (Galan et al., 1984), and, as far back as 1879, it was reported by Nasso that pale muscles of the calf change to the darker color of the adult muscle when the diet is changed from milk to plant matter (Needham, 1926). The basic problem is that veal paleness often is a combination of mild PSE with myoglobin deficiency (Swatland, 1985a). Veal color is influenced by the iron concentration of milk replacers fed in the first couple of months of fattening, and the effects are more readily monitored by color measurement of certain muscles, such as the semimembranosus, than they are by subjective examination of the carcass (Miltenburg et al., 1992). The effects of pH-related paleness and low myoglobin concentration may be difficult to separate in a practical situation, and these two factors differ in their impact between different muscles of the veal carcass (Guignot et al., 1992). Thus, the separation of these two causes of paleness in on-line measurements is desirable.

Lipid Oxidation

Myoglobin oxidation is related to lipid oxidation, which allows some useful inferences to be made from the color change of myoglobin oxidation, which is itself important and may be manipulated by vitamin E levels fed to the animal (Arnold et al., 1993). Unfortunately, exogenous vitamin E added postmortem is far less effective than endogenous vitamin E provided through feeding (Mitsumoto et al., 1993), but the enhancement of meat color by minimizing metmyoglobin in processed products may be achieved using lactic acid bacteria such as *Kurthia* and *Chromobacterium* (Arihara et al., 1993) and might conveniently be monitored on-line. In other situations, redness may be undesirable, as when *Fusarium* spoilage of feed for turkeys causes a red discoloration in otherwise pale muscles (Wu et al., 1994).

Measurement

Muscles with relatively low concentrations of myoglobin have fiber-optic reflectance spectra of correspondingly higher intensity, as seen in Figure 7.6, where the fiber-optic reflectance spectrum for washed beef has been retained for reference (spectrum 1). In Figure 7.6, spectra 2 and 3 are for diaphragm and pectoralis muscles of white veal carcasses, respectively. The spectra intersect around 640 nm. Spectrum 4 is for turkey supracoracoideus muscle. Spectra 5 and 6 are for chicken leg and breast muscle, respectively, and they intersect around 620 nm. At the present time, no critical test has been made of the intersection of the spectra for muscles differing in myoglobin concentration.

The possible low reflectance towards 700 nm of muscles with an otherwise pale appearance to the human eye is itself not a contradiction because of the low relative visibility of higher wavelengths (0.0041 at 700 nm compared to 1.0002 at 555 nm; Weast and Astle, 1979). It is important to remember that a high myoglobin concentration in a muscle is likely to be associated with features such as a high ultimate pH (because of a low initial glycogen level), a high mitochondrial density,

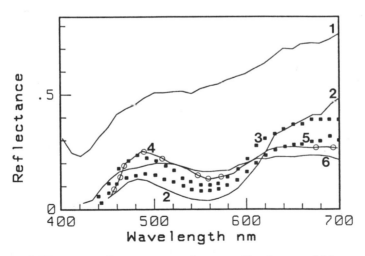

Figure 7.6 *Fiber-optic reflectance spectra for meat with a low myoglobin concentration. For reference, the fiber-optic spectrum for washed beef (1) is carried forward from spectrum 1 of Figure 7.5. Spectra 2 and 3 are for diaphragm and pectoralis muscles, respectively, of white veal carcasses (from Swatland, 1985b). Spectrum 4 is for turkey supracoracoideus muscle (from Swatland and Lutte, 1984), and spectra 5 and 6 are for leg and breast muscles, respectively, of chicken [from Swatland (1983)].*

a large amount of sarcoplasm between myofibrils, a high density of triglyceride droplets within muscle fibers, and a well-developed capillary system. Thus, microstructural scattering of light, as well as selective absorbance, may vary with the apparent redness of a muscle.

Veal color is checked on-line in Canada using a Criterion Instruments (Scarborough, Ontario) colorimeter to sort the veal into one of four color grades (Pommier, 1988). Measurements on the pectoralis give acceptable results for predicting the color of the longissimus dorsi.

Transmittance versus Reflectance

Figure 7.7 shows a set of spectra obtained from beef muscles by different methods. Spectrum 1 is a transmittance spectrum of beef fasciculi (each several fibers in depth) obtained with a microscope spectrophotometer. The transmittance was of low intensity, even with such a short path length. Spectrum 2 in Figure 7.7 is a reflectance spectrum obtained via a coherent fiber-optic light guide. A coherent light guide is one in which the relative positions of the optical fibers are maintained so that an image may be transmitted. The spectrum was measured with an airspace above the meat

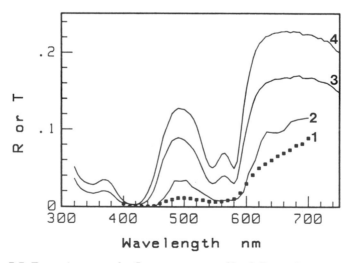

Figure 7.7 *Transmittance and reflectance spectra of beef. Transmittance spectrum 1 was measured with a microscope spectrophotometer [from Swatland (1988a)], reflectance spectrum 2 was measured with a coherent fiber-optic light guide [from Swatland (1985a)], and oxymyoglobin spectra 3 and 4 were measured with quartz optical fibers coaxial with, or perpendicular to, muscle fibers, respectively [from Swatland (1986a)].*

sample and a lens to focus the image of the sample on the light guide. Thus, the reflectance originated from a fasciculus of beef similar to that used for transmittance in spectrum 1 of Figure 7.7. Combinations of transmittance and reflectance measurements are useful in the investigation of the spectral basis of light scattering in meat (Birth et al., 1978). However, since both transmittance and reflectance change together, they must both be measured together. Hence, sample transmittance cannot logically be held as a constant in the calculation of K/S ratios from reflectance measurements (Snyder, 1968).

Inspection of Figures 7.2 to 7.5 shows there is evidently considerable variation in the shape of spectra between 580 and 700 nm. In some spectra, the increase from 580 to 700 nm is almost linear, while, in others, most of the increase occurs close to 600 nm so that spectra are strongly curved from 580 to 700 nm. These differences may originate from the formation of oxymyoglobin (compare spectra 1 and 2 in Figure 7.1). Spectra 3 and 4 in Figure 7.7 show well-defined oxymyoglobin spectra for beef measured through quartz optical fibers to allow measurements down to 320 nm (the oxymyoglobin developed during the course of time-lapse studies postmortem). Changes in the Soret absorbance bands as deoxymyoglobin is oxygenated have, so far, prevented any direct measurements of postmortem glycogenolysis in meat via fiber optics. The absorbance maximum of glycogen in water is at 420 nm (Weast and Astle, 1979). Oxymyoglobin formation would not be expected in reflectance spectra measured by fiber optics on intact carcasses, although oxymyoglobin readily develops on cut slices of meat used for macroscopic reflectance spectrophotometry (spectrum 5 in Figure 7.2 and spectra 3 to 5 in Figure 7.3).

GREEN DISCOLORATION

The discoloration of raw meat by the formation of green myoglobin derivatives such as sulfmyoglobin and cholemyoglobin (Lawrie, 1985) is very rare and only occurs in unusual circumstance. Siller and Wight (1978) briefly reported (without showing the spectra) some results obtained by photoacoustic spectroscopy of compacted disks of green muscle from turkeys with deep pectoral myopathy. Differences between normal and green muscle were found at 430 and 590 nm. Although the biochemical identity of the green pigment was not identified, it was probably a product of hemoglobin degradation in the ischaemic supracoracoideus. This pigment provides an example of the worst type of green discoloration conceivable (spectrum 1 in Figure 7.8). Meat with this general shape of spectrum appears green because of the great sensitivity of the human eye to light around 550 nm, even though there is no peak at 550 nm in the spectrum.

The iridescence of smoked pastrami has a peak at 560 nm (spectrum 2 in Figure 7.8), and the meat has a metallic green sheen. Meat iridescence originates from interference effects in the upper layers of the meat. Iridescence may be distinguished from myoglobin degradation by the fact that it disappears when light is transmitted (through a thin slice), rather than reflected, and by its angular dependency. Thus, as iridescent meat is rotated relative to the light source and the observer, the iridescence may disappear or flash to another color such as a second-order orange (Swatland, 1988). The iridescence colors of meat may be quantified using the Michel-Lévy color chart of interference colors, found in most textbooks on optical mineralogy.

ON-LINE COLOR MEASUREMENT

The implementation of on-line color difference tolerancing for meat is quite difficult if the general principles used for color control in other

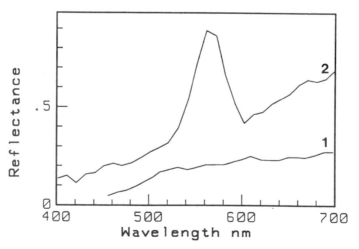

Figure 7.8 *Green discoloration of meat. Spectrum 1 is for green lesions of deep pectoral myopathy in turkeys measured with fiber optics (from Swatland and Lutte, 1984), and spectrum 2 is for green iridescence on cooked beef measured with a microscope spectrophotometer [from Swatland (1984)].*

industries are followed (Billmeyer and Saltzman, 1981), because meat has many special properties that make it difficult to assess. Assuming that an attainable industry standard for the color of a cut of meat or a product has been set in a conventional manner, the ability of the customer to see a difference between the standard and a particular sample on display is called perceptibility. The sample is acceptable if the perceptibility difference between it and the standard is less than a defined tolerance. But, usually, the standard is established on a cut meat surface using an integrating sphere colorimeter and specified in the CIE (Commission International de l'Éclairage) color space, perhaps even with a CMC (Color Measurement Committee of the Society of Dyers and Colorists) ellipsoid volume. The ellipsoid volume is more suited to human perceptibility limits than is a rectangular tolerance box, and any color within the ellipsoid centered on the standard is acceptable, while points outside the ellipsoid are unacceptable.

The special problem in the meat industry is that we would like to use an on-line probe measurement on the intact carcass or batch of meat product to predict the color of the final cut surfaces presented to the customer, and it is not realistic for us to base our color tolerancing on measurements of cut surfaces. Also, we have an extra problem, in that the color of a meat surface evolves, usually slowly but sometimes rapidly, depending on oxygen availability at different depths in the meat, refrigeration, packaging, temperature, illumination, and a host of other factors. Thus, if we adopt a conventional approach based on cut surface colorimetry, we still have the problem of placing time limits and environmental constraints on the standard. In short, a sample conforming to the ideal color standard may not appear to have an ideal color for very long.

The feasibility of obtaining CIE coordinates from probe measurements was considered earlier (Figure 5.13), and this possibility provides one route to solve the problems outlined above. But a more simple route is to work in the reverse direction to normal color differencing logic:

(*1*) Define the variation in probe readings for a large, representative population of carcasses or batches of product.

(*2*) Identify a subset of samples that represents the whole range and follow them through to the meat counter.

(*3*) Under ideal conditions of presentation (time since processing, refrigeration history, counter temperature, illumination, packaging, etc.), establish the limits to subjective acceptability, using trained panelists or representative customers to give each sample a subjective rating on the maximum point scale that seems realistic. For meat, a six-point scale such as that developed by Nakai et al. (1975) for Japanese pork is appropriate, splitting the difference for samples that are between two integers of the scale to get a functional eleven-point scale.

(*4*) Isolate a subset of samples with an ideal appearance, and follow their color degradation under typical commercial conditions.

From step (3) may be found the probe signals of acceptable and unacceptable samples. If the probe gives only a scalar (single) reading, the cutoff points for acceptability may be set immediately. If the probe gives a vector (spectrum), it may be searched using regression analysis against the quantified customer responses to find the most useful wavelengths for prediction. Multivariate analysis may be used to select a cluster of wavelengths that gives a more accurate prediction of the customer rating than a single predictor. However, it is a mistake to develop a complex prediction equation that is readily invalidated by a small change in measuring protocol. This analysis may be repeated for step (4), which may allow samples with a long shelf life to be identified. This indicates how long a sample with an ideal color may be expected to last.

REFERENCES

Arihara, K., H. Kushida, Y. Kondo, M. Itoh, J. B. Luchansky, and R. G. Cassens. 1993. *J. Food Sci.*, 58:38.

Arnold, R. N., S. C. Arp, K. K. Scheller, S. N. Williams, and D. M. Schaefer. 1993. *J. Anim. Sci.*, 71:105.

Bailey, M. E. 1968. *Proceedings of the Meat Industry Research Conference*. Chicago, IL: American Meat Institute Foundation, pp. 1–3.

Beecher, G. R., R. G. Cassens, W. G. Hoekstra, and E. J. Briskey. 1965. *J. Food Sci.*, 30:969.

Bertelsen, G. and L. H. Skibsted. 1987. *Meat Sci.*, 19:243.

Billmeyer, F. W. and M. Saltzman. 1981. *Principles of Color Technology*. New York: John Wiley & Sons.

Birth, G. S., C. E. Davis, and W. E. Townsend. 1978. *J. Anim. Sci.*, 46:639.

Bowen, W. J. 1949. *J. Biol. Chem.*, 179:235.

Broumand, H., C. O. Ball, and E. F. Stier. 1958. *Food Technol.*, 12:65.

Conway, J. M., K. H. Norriss, and C. E. Bodwell. 1984. *Am. J. Clin. Nutr.*, 40:1123.

Davis, C. E., G. S. Birth, and W. E. Townsend. 1978. *J. Anim. Sci.*, 46:634.

Dean, R. W. and C. O. Ball. 1960. *Food Technol.*, 14:271.

Elliott, R. J. 1967. *J. Sci. Food Agric.*, 18:332.

Faustman, C. and R. G. Cassens. 1991. *J. Anim. Sci.*, 69:184.

Fleming, H. P., T. N. Blumer, and H. B. Craig. 1960. *J. Anim. Sci.*, 19:1164.

Francis, F. J. and F. M. Clydesdale. 1975. *Food Colorimetry: Theory and Application.* Westport, CT: AVI Publishing Co.

Galan, P., S. Hercberg, and Y. Touitou. 1984. *Comp. Biochem. Physiol.*, 77B:647.

Guignot, F., Y. Quilichini, M. Renerre, A. Lacourt, and G. Monin. 1992. *J. Sci. Food Agric.*, 58:523.

Hall, J. E., J. M. Schwinghamer, and B. Lalone. 1976. *Am. J. Physiol.*, 230:569.

Hand, A. R. 1974. *J. Histochem. Cytochem.*, 22:207.

Hunt, M. C. 1980. *Proc. Recipr. Meat Conf.*, 33:41.

Jenkins, R. R. and J. Tengi. 1981. *Experientia*, 37:67.

Kagen, L. J. 1973. *Myoglobin. Biochemical, Physiological and Clinical Aspects.* New York: Columbia University Press.

Kastenschmidt, L. L., W. G. Hoekstra, and E. J. Briskey. 1968. *J. Food Sci.*, 33:151.

Kiese, M. and H. Kaeske, H. 1942. *Biochem. Z.*, 312:121.

Krzywicki, K. 1979. *Meat Sci.*, 3:1.

Lawrie, R. A. 1985. *Meat Science.* Oxford: Pergamon Press.

Lin, T-S. and H. O. Hultin. 1978. *J. Food Biochem.*, 2:39.

MacDougall, D. B. 1980. *J. Sci. Food Agric.*, 31:1371.

Millar, S., R. Wilson, B. W. Moss, and D. A. Ledward. 1994. *Meat Sci.*, 36:397.

Millikan, G. A. 1939. *Physiol. Rev.*, 19:503.

Miltenburg, G. A. J., T. Wensing, F. J. M. Smulders, and H. J. Breukink. 1992. *J. Anim. Sci.*, 70:2766.

Mims, F. M. 1975. *Optoelectronics.* Indianapolis, IN: H. W. Sams, p. 25.

Mitsumoto, M., R. N. Arnold, D. M. Schaefer, and R. G. Cassens. 1993. *J. Anim. Sci.*, 71:1812.

Morita, S., R. G. Cassens, and E. J. Briskey. 1969. *Stain Technol.*, 44:283.

Morton, R. A. 1975. *Biochemical Spectroscopy.* Bristol: Adam Hilger.

Nakai, H., F. Saito, T. Ikeda, S. Ando, and A. Komatsu. 1975. *Bull. Nat. Inst. Anim. Indust.*, *Chiba, Japan*, 29:69.

Needham, D. M. 1926. *Physiol. Rev.*, 6:1.

Pease, A. H. R. and C. Smith. 1965. *Anim. Prod.*, 7:273.

Pommier, S. A. 1988. *Can. Inst. Food Sci. Technol. J.*, 21:438.

Ray, G. B. and G. H. Paff. 1930. *Am. J. Physiol.*, 94:521.

Satterlee, L. D. and W. Hansmeyer. 1974. *J. Food Sci.*, 39:305.

Schirmer, R. E. and A. G. Gargus. 1986. *Amer. Lab.*, 18(12):30.

Schwimmer, S. 1981. *Source Book of Food Enzymology.* Westport, CT: AVI Publishing Co.

Siller, W. G. and P. A. L. Wight. 1978. *Avian Path.*, 7:583.

Snyder, H. E. 1965. *J. Food Sci.*, 30:457.

Snyder, H. E. 1968. *Proceedings of the Meat Industry Research Conference.* Chicago, IL: American Meat Institute Foundation, pp. 21–31.

Snyder, H. E. and D. J. Armstrong. 1967. *J. Food Sci.*, 32:241.

Solberg, M. 1970. *Can. Inst. Food Sci. Technol. J.*, 3:55.

Stewart, M. R., M. W. Zipser, and B. M. Watts. 1965. *J. Food Sci.* 30:464.

Strange, E. D., R. C. Benedict, R. E. Gugger, V. G. Metzger, and C. E. Swift. 1974. *J. Food Sci.*, 39:988.

Swatland, H. J. 1982. *J. Food Sci.*, 47:1940.

Swatland, H. J. 1983. *Poultry Sci.*, 62:957.

Swatland, H. J. 1984. *J. Food Sci.*, 49:685.

Swatland, H. J. 1985a. *J. Food Sci.*, 50:30.

Swatland, H. J. 1985b. *Histochem. J.*, 17:675.

Swatland, H. J. 1986a. *Meat Sci.*, 17:97.

Swatland, H. J. 1986b. *Can. Inst. Food Sci. Technol. J.*, 19:170.

Swatland, H. J. 1988. *J. Anim. Sci.*, 66:379.

Swatland, H. J. and G. H. Lutte. 1984. *Poultry Sci.*, 63:289.

Tappel, A. L. 1957. *Food Res.*, 22:404.

Theorell, H. and C. de Duve. 1947. *Arch. Biochem.*, 12:113.

Thomas, N. W. and M. D. Judge. 1970. *J. Agric. Sci., Camb.*, 74:241.

Warriss, P. D. 1977. *J. Sci. Food Agric.*, 28:457.

Weast, R. C. and M. J. Astle. 1979. *Handbook of Chemistry and Physics*, Boca Raton, FL: CRC Press, p. E-411.

Wolfe, S. K., D. A. Watts, and W. D. Brown. 1978. *J. Agric. Food Chem.*, 26:217.

Wu, W., D. Jerome, and R. Nagaraj. 1994. *Poultry Sci.*, 73:331.

CHAPTER 8

Prediction of Water-Holding Capacity

INTRODUCTION

FLUID LOSSES FROM meat are of tremendous concern commercially because they may cause the profitability of an enterprise to evaporate or go down the drain, both literally and figuratively. Fluid losses are difficult to see happening, and this makes them rather insidious. The factors governing how meat proteins hold on to their water or to water added to a product have been extensively investigated, for example, in Hamm's (1972) classic monograph on the colloidal chemistry of meat. His approach was that, if dried meat protein can be rehydrated experimentally in the laboratory by exposure to increasingly damp air, then three water compartments may be measured by the way in which water is taken up. Approximately 4% of the water becomes firmly bound as a monolayer around muscle proteins, another 4% is taken up as a looser second layer, and 10% of the water accumulates loosely between protein molecules.

Various terms such as water-holding capacity and water-binding capacity sometimes are used in a precise way, while sometimes they are not. Thus, for any particular report, it is essential to read the materials and methods to see what is actually being measured. There are many ways in which water in meat may be measured, and it is difficult to make meaningful comparisons from one method to another. Water-holding capacity sometimes is defined as the ability of meat to retain its own water under external influences such as centrifugation. Then, water-binding capacity may be defined as the ability of the meat to bind extra water added to a product. Water absorption or gelling capacity may be defined as the ability of meat to absorb water spontaneously from an aqueous environment.

METHODS OF MEASUREMENT

Press Method

Hamm (1972) championed the press method of quantifying meat exudate. Small meat samples (0.3 g) are pressed on a filter paper at 35 kg cm^{-2} between two plates. Upon removal after five minutes, the areas covered by the flattened meat sample and the stain from the meat sample are marked and measured. After subtracting the meat-covered area from the total stained area to obtain the wetted area, the water content may be calculated as

$$\text{mg } H_2O = \frac{\text{wetted area cm}^2 - 8.0}{0.0948}$$

Water-holding capacity is modified by pH and drops from a high pH around 10 to a low at the isoelectric point of meat proteins between pH 5.0 and 5.1. Below pH 5, a value only attained if the pH of a processed meat product is deliberately lowered, water-holding capacity starts to increase again. Water absorption follows water-holding capacity in this regard. Thus, as the pH of pork declines postmortem, its water-holding capacity decreases, and much of the water associated with muscle proteins is free to leave the muscle fiber.

Electron Microscopy and X-ray Diffraction

Electron microscopy and X-ray diffraction have been used to determine the structural basis of water-holding capacity in meat (Swatland and Belfry, 1985; Diesbourg et al., 1988; Irving et al., 1989a, 1989b; Swatland et al., 1989; Guignot et al., 1993). X-ray diffraction enables the lateral separation of myofilaments to be measured in small samples of meat. Towards the isoelectric point, thick and thin filaments in myofibrils move closer together and reduce the water space between them. Thus, as the pH declines postmortem, filaments move closer together, myofibrils shrink, and the volume of sarcoplasm increases. Eventually, muscle fibers deplete all their ATP, their membranes no longer confine the cell water, and fluid is lost from the muscle fiber and may contribute to exudate lost from the meat (Heffron and Hegarty, 1974). Unfortunately, there is no way of using X-ray diffraction as an on-line method for predicting water-holding capacity because it is an exacting technique that requires carefully dissected slips of muscles and exposure times of several hours, but it does provide a superior method of establishing the water-holding capacity of the filament matrix, giving a feature against which to validate on-line probes.

Fluid Compartments

The contribution of antemortem extracellular fluid to meat exudate is unknown, but the extracellular space in muscle is greatly increased after short periods of muscle activity (Fotedar et al., 1990). Thus, the efflux of fluid from the vasculature into the carcass muscles is of potential interest. A high volume of free water within muscles may accelerate the rate of glycolysis (Kolczak and Kraeling, 1986).

Water escapes from the spaces between muscle fiber bundles when they are cut, and the drip loss from PSE pork is increased to about 1.70% from a normal value of about 0.77% of trimmed carcass weight (Smith and Lesser, 1979). In sliced pork, drip losses increase with storage, from 9% after one day to 12.3% after six days (Lundström et al., 1977). This creates a serious weight loss from the carcass. In the United States, Hall (1972) combined an estimated incidence of PSE carcasses (18%) and their typical extra shrink loss (5 to 6%) to estimate total national losses as 94 to 95 million dollars. Obviously, this is only a rough guess, but it serves to emphasize the commercial importance of pH in pork. Kauffman et al. (1978) estimated that the excess weight loss from PSE pork during transport, storage, and processing could exceed 1 million kilograms per annum in the United States. The exudate from PSE pork is wasted in most countries, but, in Japan, it has been used successfully in manufacturing sausages (Tsukamasa et al., 1992).

Exudate-filled spaces between muscle fiber bundles contribute to the soft texture and easily separated fiber bundles of PSE pork because the amount of water bound within muscle fibers has an effect on meat tenderness (Currie and Wolfe, 1980). X-ray diffraction results indicate that the detachment of myosin molecule heads also may contribute to the softness of PSE pork (Irving et al., 1989a).

MONOCHROMATIC MEASUREMENTS

The subjective paleness of meat, especially pork, is often used as a guide to water-holding capacity, assuming that paleness, softness, and exudation are perfectly correlated, which is not true. Paleness and water-holding capacity are different physical phenomena: although they may share a common cause, they are partly independent, and this limits the accuracy with which water-holding capacity may be predicted from paleness (van Laack et al., 1994). However, subjective assessments of paleness are mediated via the eye, acting as a photometer, and the eye does not respond equally to all wavelengths. Light at 400 and 700 nm produces little response, while light at 550 nm produces a strong response. Objective measurements of paleness using any of the colorimetry systems will also take into account this spectral response of the human eye. Thus, paleness is weighted heavily towards 550 nm, and events at 400 and 700 nm are almost completely missed. Thus, optical properties of meat based on the whole spectrum may have a higher information content than paleness.

Reflectance

The search for on-line methods to predict water-holding capacity started in the late 1930s, but it was not until the 1960s that German researchers started to exploit the correlation of paleness with water-holding capacity or had access to commercially available handheld reflectance meters for measuring paleness. A general feature of most of these instruments is that they are, or were, monochromatic, making a reflectance measurement at a single color, usually a broad band of wavelengths.

As research progressed, there was little experimental work to discover how low pH causes paleness, even though this period was marked by great progress in understanding how low pH decreases water-holding capacity. With a variety of photometers in use, each giving a different scale of readings and many of them not originally designed for measuring meat, common trends among the results were difficult to establish.

Line 1 in Figure 8.1 shows results from MacDougall and Disney (1967) which were obtained with an Optica CF4DR instrument measuring reflectance at 525 nm. Line 2 in Figure 8.1 shows data from Barton-Gade (1981), and line 3 shows data from Lundström et al. (1979), both using a similar instrument (Zeiss Elrepho) at 535 nm. Line 4 in Figure 8.1 (Schwörer et al., 1980) was obtained with an EEL 11 (Evans Electroselenium Ltd.) reflectometer and a green filter (530 to 540 nm). These reflectance values vary when plotted against pH_1,

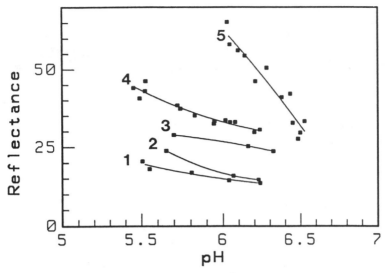

Figure 8.1 *Relationships of pH$_1$ with monochromatic reflectance for longissimus dorsi muscles of Pietrain, Pietrain × Landrace, and Landrace pigs using an Optica CF4DR [line 1, data from MacDougall and Disney (1967)]; for Danish Landrace pigs using a Zeiss Elrepho [line 2, data from Barton-Gade (1981)]; longissimus dorsi muscles of Swedish Landrace and Yorkshire pigs using a Zeiss Elrepho [line 3, data from Lundström et al. (1979)]; for longissimus dorsi muscles of Swiss Large White and Landrace pigs using an EEL 11 with green filter [line 4, data from Schwörer (1982)]; and for longissimus dorsi muscles of English commercial pigs using an EEL smoke-stain photometer [line 5, data from Evans et al. (1978) and Kempster et al. (1984)].*

probably because of differences in wavelength and method of standardization. The EEL reflectometer was first used by MacDougall et al. (1969), but the data of line 5 in Figure 8.1 are from Evans et al. (1978) and Kempster et al. (1984). The scatter plot of data shown in Figure 8.1 is for mean values (rather than raw data), and the relationship between reflectance and pH$_1$ appears to be stronger than it actually is. Kempster et al. (1984) found that reflectance may be only poorly correlated with pH$_1$ ($r = -0.17$, $r^2 = 2.9\%$), but stronger correlations of reflectance with pH$_1$ may be found in small studies under laboratory conditions (MacDougall and Disney, 1967). Essentially, all five lines shown in Figure 8.1 are fairly close to being straight-line relationships, which contrasts to the marked curvilinearity of the relationships considered next and to the results of Warris and Brown (1987), who found a biphasic relationship between drip loss and reflectance. Below pH 6.1, decreasing the pH$_{45}$ had a small effect on drip loss and a large effect on reflectance, while effects were reversed above pH 6.1.

Göfo Meter

The Göfo reflectance photometer (Ernst Schütt, Göttingen, Germany) is or was used in many countries to measure pork paleness. The Göfo is standardized against a dark plate instead of a white one so that the current to the Göfo ammeter is directly proportional to pH_l (Figure 8.2); therefore, a low Göfo value indicates pale pork. At first, it appeared that Göfo mean values might be related to pH_l by a line similar to that derived from the first equation shown below. But this was not the case in larger data sets (Schepers, personal communication) where the envelope enclosing the experimental points may vary by as much as ± 15 Göfo units from a line of best fit (line 3 in Figure 8.2). Göfo mean values appear to be related quasilinearly with $[H^+]$ rather than with pH_l. From lines 1 and 3 of Figure 8.2, respectively:

$$\text{amps} \times 10^{-6} = 70.9 - \text{antilog}_{10}(6.95 - pH_l)$$

$$\text{amps} \times 10^{-6} = 77.1 - \text{antilog}_{10}(7.073 - pH_l)$$

A similar, but reversed, equation applies to the reflectance data

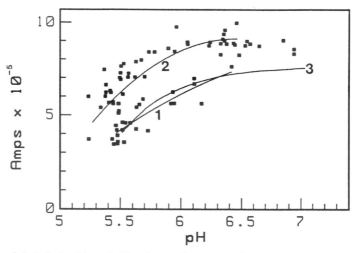

Figure 8.2 *Relationships of pH and monochromatic reflectance measured with a Göfo photometer for longissimus dorsi muscles of German Large White, Landrace, and commercial pigs relative to pH₁ [line 1 with scatter plot, data from Blendl and Puff (1978) and Schmitten et al. (1981, 1984)]; Canadian Landrace, Yorkshire, and Duroc pigs relative to pH from forty-five minutes to twenty-four hours postmortem [line 2 with scatter plot, data from Swatland (1981)]; and German pigs relative to pH₁ (line 3 only, data from Schepers, personal communication).*

obtained by MacDougall et al. (1969) with a smoke-stain reflectometer.

$$\text{reflectance} = 39.8 + \text{antilog}_{10}\,(7.02 - pH_{90})$$

The scatter in all these equations equals or exceeds that found by Schepers.

Correlations of monochromatic reflectance with pH_{90} or with pH_1, based on straight-line regressions, seldom are strong enough to be of any industrial value [details given by Bendall and Swatland (1988)]. Values for r^2 only exceed 45% in two small and one larger study, although slightly stronger relationships might exist with a polynomial fit. MacDougall and Disney (1967) and Barton-Gade (1981) found that reflectance at 525 to 535 nm is affected by myoglobin, which was and still is a problem in attempting to predict water-holding capacity from light scattering. Barton-Gade (1981) estimated that reflectance falls by about 5% for each extra 8 mg of total heme iron and found that reflectance increases by about the same amount for an extra 1.7% of fat. The pigment effect may not affect reflectance at 640 nm (Eikelenboom et al., 1974). The effect of myoglobin on Göfo reflectance probably contributed to the differences between lines 1 and 3 versus line 2 in Figure 8.2 since line 2 included many slow-growing animals with high myoglobin levels (as well as including measurements made perpendicularly to the axes of the muscle fibers).

Fiber-Optic Probe

A monochromatic fiber-optic probe was developed by MacDougall and Jones (1975) to measure the internal reflectance of meat. It has a peak sensitivity at 900 nm, where absorbance by red heme pigments is minimal. Bendall and Fentz (Bendall, personal communication) made fiber-optic reflectance measurements during the development of rigor in longissimus dorsi muscles of halothane-negative Large White pigs, near to where samples had been taken for measurements of pH and other biochemical parameters. These results (line 1 of Figure 8.3) show that fiber-optic reflectance increases slowly as the pH falls from 7 to 6.3 and then more rapidly from there down to pH 5.4, where it reaches a maximum value.

The change in fiber-optic reflectance caused by the decline in pH is described by the equation:

$$\text{fiber-optic reflectance} = 11.2 + \text{antilog}\,(6.82 - pH)$$

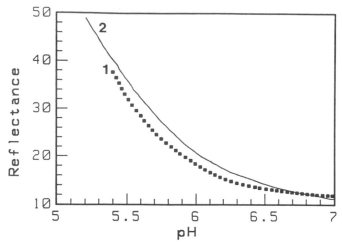

Figure 8.3 *Relationships of pH and monochromatic fiber-optic reflectance as pH falls postmortem in seven Large White carcasses (line 1, Bendall, personal communication) and for pH$_u$ of longissimus dorsi muscles of 1584 Irish pigs [line 2, Somers et al. (1985)].*

None of the muscles in these experiments had a pH$_1$ of less than 6.3, and it is extremely unlikely that protein denaturation was the cause of the increase in fiber-optic reflectance towards the ultimate pH of approximately 5.4. Somers et al. (1985; line 2 of Figure 8.3) showed that the relationship of fiber-optic reflectance with pH$_u$ was similar to that of fiber-optic reflectance with declining pH.

A large increase in fiber-optic reflectance occurs in PSS pigs with pH$_1$ <5.8 and a pH$_1$ index >60% (Cheah et al., 1984). In PSS pigs, fiber-optic reflectance in the longissimus dorsi muscles at twenty-four hours postmortem (pH$_u$ ≈ 5.4) rises to 53.4, compared with 32.9 for halothane-negative pigs, and 34 for the Large White pigs at comparable pH$_u$ values of 5.4 to 5.5 (Bendall, personal communication). Protein denaturation may occur in PSS pigs but is unlikely to be the cause of paleness in halothane-negative pigs (Penny, 1969).

Summary

In summary, research with monochromatic on-line systems shows that light scattering may be used to assess the level of PSE in pork and, indirectly, the water-holding capacity of the meat, but the latter relationship is not particularly reliable. Two main problems are the confounding effect of myoglobin and possible nonlinearity between reflectance and water-holding capacity. With regard to the mechanism of pH-related light scattering, the early research shows that it involves something in addition to protein precipitation. For example, differences in protein denaturation are unlikely to explain the difference between normal and DFD meat.

EFFECT OF pH ON WATER-HOLDING CAPACITY

If light scattering is to be used to predict water-holding capacity by exploiting the fact that both are determined by pH, then it is necessary to understand the relationship between pH and water-holding capacity, but this requires a clear understanding of how water-holding capacity is to be measured. There are many different protocols for measuring water-holding capacity, but the majority is based on only a few basic possibilities: (1) squeezing out fluid with a press, (2) centrifugation, and (3) weight loss by gravity acting on sliced muscles. Water-holding capacity generally decreases if the pH is low or the rate of pH decline is rapid, but the relationship is not exactly linear, and, with many methods, there seems to be a point of inflection at around pH 6.1. The details are given by Bendall and Swatland (1988, pp. 101 – 109), careful reading of which reveals a problem between results obtained with different methods.

Bag Drip

Bag drip methods using gravity weight loss from meat slices in an inflated bag to control evaporation (Figure 8.4, lines 1 to 3) have a marked inflection at pH 6.1, so that fluid losses increase rapidly as the pH drops from 7 to 6.1, but show little increase below pH 6.1. This may be seen in some press methods where the loose fluid is measured by filter paper area (Figure 8.5, lines 1 and 2), while others lack an inflection and have a linear relationship to pH over a wide range (Figure 8.5, lines 3 to 5). These latter data are compatible with relationships of fibrillar centrifugation water-holding capacity with pH (Figure 8.6, lines 2 and 3) but not when there is a point of inflection (Figure 8.6, line 1). Thus, bag drip increases rapidly from pH 7 down to 6.1, then has little further increase at lower pH's, whereas fibrillar water-holding capacity may have the opposite pattern, showing little change from pH 7 down to 6.1, then a rapid decrease at lower pH values.

Whatever the cause of this conflict, it creates problems for an on-line

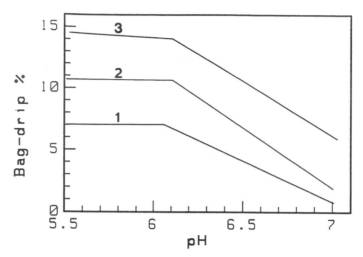

Figure 8.4 *Relationships of pH₁ with bag-drip percent for boneless pork chops at one day postmortem [line 1, data from Cheah et al. (1984)], for whole chops at one day postmortem (line 2, Large White and Landrace × Large White pigs with pH₁ > 6.0, Bendall, personal communication), and for whole chops at three days postmortem [line 3, data from Warriss (1982)].*

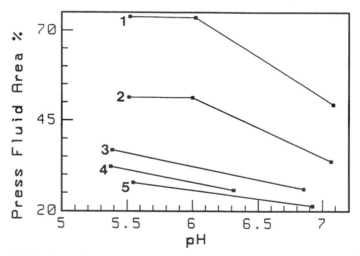

Figure 8.5 *Relationships of pH with various pressed fluid methods versus pH₁ in longissimus dorsi muscles of English Large White pigs (line 1, Bendall, personal communication), versus pH₁ in longissimus dorsi muscle of Danish Landrace pigs [line 2, Wismer-Pedersen (1959)], versus pH₁ [line 3, Scheper (1975)], versus pH₁ for Swiss Landrace and Large White pigs [line 4, Schwörer (1982)], and versus pHᵤ [line 5, Scheper (1975)].*

216

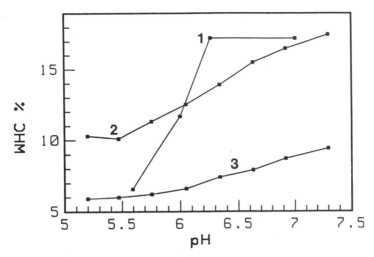

Figure 8.6 *Relationship of pH with myofibrillar water-holding capacity [line 1, Penny (1969), for muscles with pH₁ 7.0 to 5.4 homogenized in pH 7 buffer; line 2, Bendall and Wismer-Pedersen (1962), for normal muscles with pH₁ 6.3 adjusted with NaOH or HCl to pH 7.2 to 5.2; and line 3, Bendall and Wismer-Pedersen (1962), for muscles from PSS pigs with pH₁ 5.6 with pH adjusted as in line 2].*

method that attempts to predict water-holding capacity via a mutual dependency of both light scattering and water-holding capacity on pH. Low-angle X-ray diffraction patterns (Irving et al., 1989b), as seen in Figure 8.7, follow a similar pattern to that for water-holding capacity in Figure 8.6, line 1, although the inflection is at pH 6.5, rather than at pH 6.1. Reference to the error bars in the original publication (Irving et al., 1989b) shows that the point of inflection is real and unlikely to be caused by chance, and a similar effect has been observed in muscles of other species. But why is the major component of bag drip loss above pH 6.1 (Figure 8.3) if fibrils do not start to release fluid from between their filaments until below pH 6.5 (Figure 8.7)?

Source of Fluid

To answer this question requires some consideration of downstream fluid compartments from the primary source of fluid from within the filament lattice. Measuring filament separation by X-ray diffraction and the interfiber space by differential interference contrast microscopy, at approximately constant sarcomere lengths measured by laser diffraction, and assuming from stereological grounds that these areas are proportional to the volumes of their corresponding fluid compartments,

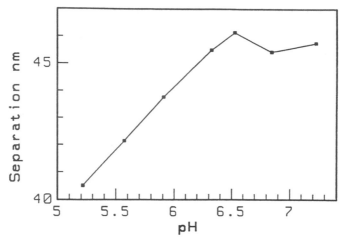

Figure 8.7 *Effect of pH on the lateral separation of filaments in pork longissimus dorsi measured by low-angle X-ray diffraction (Irving et al., 1989b).*

then filament lattice spacing (converted to lattice area to be comparable with other areas) is negatively correlated with interfiber area ($r = -.67$, $P < .01$, Figure 8.8), and interfiber area is negatively correlated with centrifugation water-holding capacity ($r = -.75$, P < .005; Figure 8.9). A volume measurement (water-holding capacity) should not really be plotted against an area, and the correlation is slightly stronger ($r = -0.78$) if this is corrected; however, no significant relationship ($P > .05$) yet has been detected between filament lattice area and water-holding capacity ($r = .28$), which demands an explanation.

The primary reservoir of releasable muscle fluid from pork is probably from within the filament lattice of the live stress-resistant pig, although PSS pigs might have some other compartment that is unusually swollen, ultimately to release fluid from the meat. But in rat extensor digitorum longus (Zierler, 1973), the total water content (0.80 ml g^{-1} wet muscle weight) is partitioned between extracellular space (0.13 ml g^{-1}), sarcoplasmic reticulum volume (0.13 ml g^{-1}) and sarcoplasmic volume (0.54 ml g^{-1}). By extrapolation to zero time postmortem, using the data from Swatland and Belfry (1985) for porcine longissimus dorsi, it seems that 20% or less of the sarcoplasmic volume is interfibrillar space (if rat and porcine muscles are comparable) and that the minimum intrafibrillar (interfilament) space is in the order of 0.43 ml g^{-1}. Corrections are needed for variations in intramuscular fat content, but the interfilament space looks like the major primary reservoir of fluid released from pork

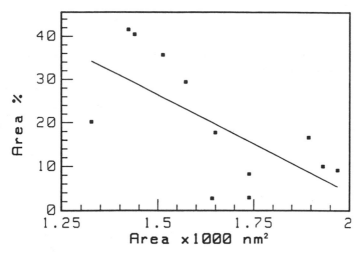

Figure 8.8 *Relationship in pork loins of myofilament lattice area measured by X-ray diffraction with the area between muscle fibers measured by differential interference contrast microscopy (Swatland et al., 1989).*

as a result of a postmortem decline in pH. The interfilament compartment is highly dependent on pH (Diesbourg et al., 1988), whereas a pH dependency has yet to be demonstrated for the other compartments but could exist. The sarcoplasm between the fibrils and the extracellular fluid between the fibers also may contribute some fluid to be released,

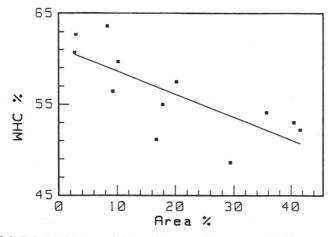

Figure 8.9 *Relationship in pork loins of the area between muscle fibers measured by differential interference contrast microscopy with centrifugation water-holding capacity percent (Swatland et al., 1989).*

but the relative amount is unknown. Swollen remains of mitochondria and sarcoplasmic reticulum in the interfibrillar compartment might conceivably trap fluid, rather than release it, as a consequence of a postmortem decline in pH. Thus, when interfilament area is negatively correlated with interfiber area (Figure 8.8), perhaps much of the fluid released from the filament lattice has been transferred to, and is still present in, the interfiber space at the time of sampling (when frozen in liquid nitrogen).

Likewise, the interfiber and interfascicular space, rather than the interfibrillar space, could be the reservoir from which fluid may be extracted by high-speed centrifugation. Centrifugation causes a decrease in filament separation of only 2.6 nm ($P < .01$), corresponding to a 12% decrease in filament lattice volume. This amounts to a decrease in interfilament space of less than 14%, whereas fluid losses from centrifugation may range from 17 to 49%. In other words, centrifugation pulls out fluid from other fluid compartments, in addition to that from the filament lattice of fibrils, and the decrease in filament lattice area caused by centrifugation is only weakly correlated with water-holding capacity ($r = .22$). Thus, differences between samples in water-holding capacity may be caused by compartments downstream from the filament lattice, possibly the reservoir between fibrils and demonstrably the reservoir between fibers. Thus, the relationship between filament separation and water-holding capacity is indirect. The only relationship between filament lattice separation and water-holding capacity might be via their mutual relationship with interfiber space.

Although the filament lattice probably is the major source of meat exudate, under typical commercial conditions, much of its fluid already may have been lost and may be awaiting release from downstream compartments. X-ray diffraction measurements on a fairly consistent population of Ontario pigs show a loss of fluid from the filament lattice in the first few hours after slaughter, with little subsequent change, and perhaps even a slight re-uptake of fluid (Figure 8.10). By about six hours postmortem, transmission electron microscopy shows that the released fluid has swollen the sarcoplasmic compartment between the fibrils to about its maximum (Figure 8.11). In summary, the spatial sequence of fluid release in normal pigs is from the filament lattice to the interfibrillar sarcoplasm to the interfiber space, and, finally, as exudate lost through the interfascicular space. Thus, apparently conflicting data relationships seen in Figures 8.4 to 8.7 could be an illusion. Consider the analogy of rain-bearing clouds in a major storm, first creating a deluge high in a

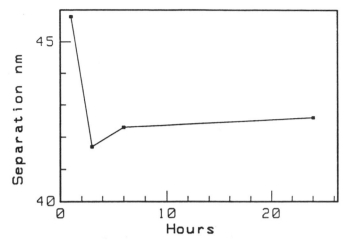

Figure 8.10 *Decrease in lateral separation between myofilaments measured by X-ray diffraction in pork loins postmortem (Diesbourg et al., 1988).*

watershed and then swelling the volume of wetlands and rivers. To estimate the flood risk down on the delta, we need to look at the pattern of the water volume moving downstream, and instantaneous correlations of precipitation on the hills with flooding in the delta might be weak or zero.

In summary, the source of exudate from pork is traceable by sequential

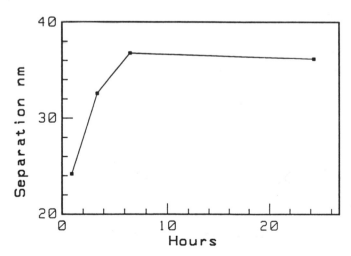

Figure 8.11 *Increase in the sarcoplasmic area between myofibrils measured by electron microscopy in pork loins postmortem (Swatland and Belfry, 1985).*

relationships down the pathway from the filament lattice to the meat surface, but direct correlations of the first and last compartments may be very weak if no account is made of the lapse of time. In other words, looking at a real-time effect of pH (e.g., changing the pH of samples for X-ray diffraction) is not the same as examining the correlation of drip loss with ultimate pH.

PREDICTION BY SPECTROPHOTOMETRY

The use of optical fibers to obtain reflectance measurements from within a sample of meat, pioneered by MacDougall and Disney (1967), was a logical continuation of earlier work using glass light pipes with an internally reflecting silvered surface. Spectrophotometry through optical fibers, instead of monochromatic measurements, was the next logical step in the development of this technology (Swatland, 1982). Meat spectrophotometry goes back a long way, at least in Canada (Winkler, 1939), but recent advances in optoelectronics were required before it was possible to make handheld instrumentation for industrial use. Theoretically, there are several advantages of spectrophotometry over monochromatic measurements:

(1) The possibility of measuring myoglobin separately from pH-related light scattering, thus removing the confounding effects of age-related myoglobin discovered by MacDougall and Disney (1967) and Barton-Gade (1981), which is particularly important in grading veal paleness (Swatland, 1985a)

(2) The possibility of using different regions of the spectrum to build a compound prediction equation for water-holding capacity

(3) The possibility of using different regions of the spectrum to make other measurements of value in meat science, such as the yellowness (Swatland, 1988a) or softness of fat (Irie and Swatland, 1992a)

Despite the potential advantages of spectrophotometry, monochromatic measurements are far easier because monochromatic apparatus have fewer components and less complexity than spectrophotometric apparatus, thus reducing the cost and increasing the durability. For example, the Danish MQM fat-depth probe (Figure 3.8) also provides a commercially satisfactory measure of water-holding capacity (Andersen et al., 1993).

Colormet and Other Fiber-Optic Spectrograph Probes

The Canadian Colormet meat probe (Swatland, 1986) is a multipurpose reflectance spectrophotometer (400 to 700 nm) developed to measure a number of parameters of interest in meat science, using a xenon flash unit synchronized with a photodiode-array spectrometer and connected through a stainless steel fiber-optic probe. Similar instruments are available in Denmark (Andersen et al., 1993) and Japan (Tsuruga et al., 1994).

Results obtained with the Colormet meat probe explain some of the conflicting results obtained earlier with monochromatic devices, because different regions of the spectrum may be seen to have different relationships with pH and with filament separation. Figure 8.12 shows the spectral distribution of correlation coefficients of longissimus dorsi reflectance with gluteus medius reflectance (line 1), of longissimus dorsi reflectance with pH_u (line 2), and the absolute value of r (which is negative) for the correlation of longissimus dorsi reflectance with X-ray diffraction measurements of interfilament spacing (line 3). Interfilament

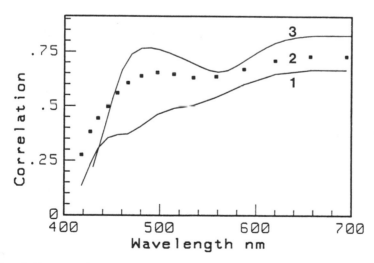

Figure 8.12 *Spectral distribution of absolute values for correlation coefficients of fiber-optic reflectance in longissimus dorsi with fiber-optic reflectance in gluteus medius [line 1, Swatland (1986)], with longissimus dorsi pH_u [line 2, Swatland (1986)], and with longissimus dorsi filament separation measured by X-ray diffraction (line 3, negative correlation from unpublished data of Irving, Swatland, and Millman). All samples were from commercial pork loins with a range from mild PSE to mild DFD.*

spacing and light scattering are strongly correlated. In this example, where the optical measurement was internal reflectance measured via optical fibers, reflectance of red light provides a good measure of the pH-dependent state of the filament lattice with the least error originating from intermuscular variation. Low correlations around 555 nm in lines 2 and 3 probably originate from variance in myoglobin concentration between carcasses. For Figure 8.12, the range in pork quality was from mild PSE to mild DFD and was representative of commercial pork from halothane-negative pigs. Stronger correlations might have been obtained if a greater range for PSE to DFD had been used.

The rate at which light scattering increases after slaughter is important because it determines when measurements may be made and with what reliability. When optical apparatus to measure pork paleness is maintained at a constant position on a static carcass (so as to avoid sampling error), paleness may show a transient decrease before its ultimate increase (Figure 8.13; Swatland, 1985b). Given that the time of occurrence of the reflectance minimum differs between animals, this is a major problem in the prediction of ultimate water-holding capacity from optical measurements made soon after slaughter. Muscle fibers initially may take up extracellular fluid as their internal osmotic pressure rises at the start of postmortem glycolysis.

Some typical results for on-line prediction of water-holding capacity of unfrozen, frozen and thawed, and thin-sliced pork are shown in Figure

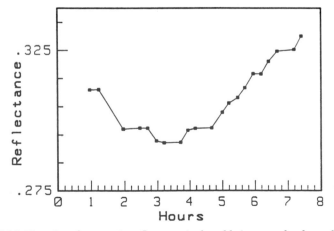

Figure 8.13 *Transient decrease in reflectance in the adductor muscle of a pork carcass postmortem (Swatland, 1985b).*

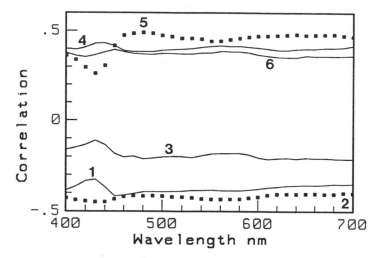

Figure 8.14 *Correlations of fiber-optic reflectance with water-holding capacity of unfrozen pork (line 1), water-holding capacity after freezing and thawing (line 2), slicing losses of unfrozen pork (line 3), slicing losses after freezing and thawing (line 4), drip loss in unfrozen pork (line 5), and drip loss after freezing and thawing (line 6) [from Irie and Swatland (1993)].*

8.14. Only simple correlations are shown, and stronger prediction relations can be obtained using a combination of wavelengths, but this should only be undertaken once apparatus and methodology have been standardized. Choice of wavelengths depends on the population of pigs to be measured (Swatland, 1988b). The use of wavelength-position matrices to enhance the prediction of water-holding capacity was considered earlier (Table 5.1).

PREDICTION BY CAPILLARY METHODS

Hofmann (1975) used capillary action to assess the abundance of free fluid on meat. Fluid drawn into one side of a ceramic disk displaces air and can be measured by the change in pressure in an enclosed space on the other side of the disk. However, the method was not rated as reliable for the prediction of water-holding capacity in pork by Kauffman et al. (1986). The tensiometer method, where meat juice replaces water rather than air, has similar problems in that capillary surfaces are easily blocked, especially if repeat use is attempted (Kim et al., 1995).

REFERENCES

Andersen, J. R., C. Borggaard, T. Nielsen, and P. A. Barton-Gade. 1993. *39th International Congress of Meat Science and Technology.* Calgary, Alberta, pp. 153–164.

Barton-Gade, P. A. 1981. The measurement of meat quality in pigs post mortem, in *Porcine Stress and Meat Quality,* T. Frøystein, E. Slinde, and N. Standal, eds., As, Norway: Agricultural and Food Research Society, pp. 205–218.

Bendall, J. R. and H. J. Swatland. 1988. *Meat Sci.,* 24:85.

Bendall, J. R. and J. Wismer-Pedersen. 1962. *J. Food Sci.,* 27:144.

Blendl, H. M. and H. Puff. 1978. *Fleischwirtshaft,* 58:1702.

Cheah, K. S., A. M. Cheah, A. R. Crosland, J. C. Casey, and A. J. Webb. 1984. *Meat Sci.,* 10:117.

Currie, R. W. and F. H. Wolfe. 1980. *Meat Sci.,* 4:123.

Diesbourg, L., H. J. Swatland, and B. M. Millman. 1988. *J. Anim. Sci.,* 66:1048.

Eikelenboom, G., D. R. Campion, R. G. Kauffman, and R. G. Cassens. 1974. *J. Anim. Sci.,* 39:303.

Evans, D. G., A. J. Kempster, and D. E. Steane. 1978. *Livestock Prod. Sci.,* 5:265.

Fotedar, L. K., J. M. Slopis, P. A. Narayana, M. J. Fenstermacher, J. Pivarnik, and I. J. Butler. 1990. *J. Appl. Physiol.,* 69:1695.

Guignot, F., X. Vignon, and G. Monin. 1993. *Meat Sci.,* 33:333.

Hall, J. T. 1972. Economic importance of pork quality, in *The Proceedings of the Pork Quality Symposium,* R. Cassens, F. Giesler, and Q. Kolb, eds., Madison, WI: University of Wisconsin Press, pp. ix–xii.

Hamm, R. 1960. *Adv. Food Res.,* 10:355.

Hamm, R. 1972. *Kolloidchemie des Fleisches—das Wasserbindungsvermoegen des Muskeleiweisses in Theorie und Praxis.* Berlin: Paul Parey.

Heffron, J. J. A. and P. V. J. Hegarty. 1974. *Comp. Biochem. Physiol.,* 49A:43.

Hofmann, K. 1975. *Fleischwirtschaft* 55:25.

Irie, M. and H. J. Swatland. 1992a. *Asian Austral. J. Anim. Sci.,* 5:753.

Irie, M. and H. J. Swatland. 1992b. *Food Res. Internat.,* 25:21.

Irie, M. and H. J. Swatland. 1993. *Meat Sci.,* 33:277.

Irving, T. C., H. J. Swatland, and B. M. Millman. 1989a. *J. Anim. Sci.,* 67:152.

Irving, T. C., H. J. Swatland, and B. M. Millman. 1989b. *Can. Inst. Food Sci. Technol. J.,* 23: 79–81.

Kauffman, R. G., D. Wachholz, D. Henderson, and J. V. Lochner. 1978. *J. Anim. Sci.,* 46:1236.

Kauffman, R. G., G. Eikelenboom, P. G. van der Wal, B. Engle and M. Zaar. 1986. *Meat Sci.* 18:307.

Kempster, A. J., D. G. Evans, and J. P. Chadwick. 1984. *Anim. Prod.,* 39:455.

Kim, B. C., R. G. Kauffman, J. M. Norman and S. T. Joo. 1995. *Meat Sci.* 39:363.

Kolczak, T. and R. R. Kraeling. 1986. *J. Anim. Sci.,* 62:646.

Lundström, K., H. Nilsson, and I. Hansson. 1977. *Swedish J. Agric. Res.,* 7:193.

Lundström, K., H. Nilsson, and B. Malmfors. 1979. *Acta Agric. Scand. Suppl.,* 21:71.

MacDougall, D. B. and J. G. Disney. 1967. *J. Food Technol.,* 2:285.

MacDougall, D. B. and S. J. Jones. 1975. *J. Sci. Food Agric.,* 31:1371.

MacDougall, D. B., A. Cuthbertson, and R. J. Smith. 1969. *Anim. Prod.*, 11:243.

Penny, I. F. 1969. *J. Food Technol.*, 4:269.

Scheper, J. 1975. *Fleischwirtshaft*, 55:1176.

Schmitten, F., K-H. Schepers, E. Wagner, and W. Trappmann. 1981. *Schweinen. Zuchtungskunde*, 53:125.

Schmitten, F., K-H. Schepers, H. Jüngst, U. Reul, and A. Festerling. 1984. *Fleischwirtshaft*, 64:1238.

Schwörer, D. 1982. Doctor's Thesis, ETH. 6978, Tech. Hochschule, Zurich, pp. 1–138.

Schwörer, D., J. K. Blum, and A. Rebsamen. 1980. *Livestock Prod. Sci.*, 7:337.

Smith, W. C. and D. Lesser. 1979. *Anim. Prod.*, 28:442.

Somers, C., P. V. Tarrant, and J. Sherington. 1985. *Meat Sci.*, 15:63.

Swatland, H. J. 1981. *Can. Inst. Food Sci. Technol. J.*, 14:147.

Swatland, H. J. 1982. *J. Food Sci.*, 47:1940.

Swatland, H. J. 1985a. *J. Food Sci.*, 50:1489.

Swatland, H. J. 1985b. *J. Anim. Sci.*, 61:887.

Swatland, H. J. 1986. *Can. Inst. Food Sci. Technol. J.*, 19:170.

Swatland, H. J. 1988a. *J. Sci. Food Agric.*, 46:195.

Swatland, H. J. 1988b. *Can. Inst. Food Sci. Technol. J.*, 21:494.

Swatland, H. J. 1988c. *J. Anim. Sci.*, 66:2578.

Swatland, H. J. and S. Belfry. 1985. *Mikroskopie*, 42:26.

Swatland, H. J. and M. Irie. 1992. *J. Anim. Sci.*, 70:2138.

Swatland, H. J., T. C. Irving, and B. M. Millman. 1989. *J. Anim. Sci.*, 67:1465.

Tsukamasa, Y., K. Fukumoto, M. Ichinomiya, M. Sugiyama, Y. Minegishi, Y. Akahane, and K. Yasumoto. 1992. *Nippon Shokuhin Kogyo Gakkaishi*, 39:862.

Tsuruga, T., T. Ito, M. Kanda, S. Niwa, T. Kitazaki, T. Okugawa, and S. Hatao. 1994. *Meat Sci.*, 36:423.

van Laack, R. L. J. M., R. G. Kauffman, W. Sybesma, F. J. M. Smulders, G. Eikelenboom, and J. C. Pinheiro. 1994. *Meat Sci.*, 38:193.

Warris, P. D. 1982. *J. Food Technol.*, 17:573.

Warriss, P. D. and S. N. Brown. 1987. *Meat Sci.*, 20:65.

Winkler, C. A. 1939. *Canad. J. Res. D*, 17:1.

Wismer-Pedersen, J. 1959. *Food Res.*, 24:711.

Zierler, K. L. 1973. Some aspects of the biophysics of muscle, in *The Structure and Function of Muscle, Vol. III*, G. H. Bourne, ed., New York: Academic Press, pp. 117–183.

CHAPTER 9

Connective Tissue Fluorescence

INTRODUCTION

THE IMPORTANCE OF connective tissue in meat toughness is a topic for lively debate among meat scientists. We all agree that connective tissues are responsible for major differences in tenderness between cuts of meat, such as in the contrast of stewing beef with prime beef steak, and for differences in tenderness between beef from young and old cattle. But there is considerable disagreement about the role of connective tissues in the real world — when customers complain about toughness in a prime roast or barbecue steak (McDonell, 1988). Those who dispute the importance of connective tissue argue that, despite the wealth of biochemical information published on connective tissue in meat (Bailey and Light, 1989), little has been published that correlates collagen biochemistry with taste panel responses to commercial beef. Refereed journals seldom allow negative findings to be reported because failure to find an effect does not necessarily constitute a proof of nonexistence. Thus, we do not know how many research groups have searched for such relationships and failed.

Given these uncertainties, it has become popular to argue that connective tissues are gelatinized by appropriate cooking methods so that any toughness in a prime steak must originate from the state of the myofibrils or cytoskeleton, leading to the conclusion that connective tissues do not play a major role in commercial problems with meat toughness (Dransfield, 1992). Superimposed on this difficult situation is the fact that, when some researchers talk of meat toughness, they are talking only of beef toughness, while others are referring primarily to another species, such as lamb, pork, or poultry. In lamb, collagen concentration is correlated with sensory panel results (Young and Braggins, 1993), but the situation for U.S. beef still is uncertain.

There is no doubt whatsoever that sarcomere length and integrity, plus the state of the cytoskeleton, are very important in explaining meat toughness in the real world, as in the notorious toughness of cold-shortened meat and beef consumed before aging. These topics are given only passing attention in this book, simply because we have not started to measure them on-line and not because these topics are unimportant.

This chapter describes the current state of fluorescence technology for measuring connective tissue in meat on-line. Readers should be aware, however, that the current weight of opinion in the United States and elsewhere is against connective tissue being a major cause of toughness in prime beef; however, the weight of opinion sometimes swings like a pendulum. Connective tissue fluorometry either will be useful or it will not, and it is too early yet to make the final judgement. Ultimately, even if the technology cannot improve quality control of the tenderness of prime beef, it has a stronger chance of identifying mature beef that is still tender. This offers an immediate commercial advantage, saving valuable meat from carcasses downgraded by age.

There is sufficient information on connective tissue fluorescence to justify a chapter, but it is important to note that fluorescence is not the only on-line method available for collagen. Andersen et al. (1993) developed a NIR fiber-optic probe that gives an accurate prediction of collagen over a wide range of beef carcasses, $R^2 = 0.91$.

CONNECTIVE TISSUE BIOLOGY

The fibrous connective tissues in meat form a continuous meshwork, starting with microscopic strands of endomysium around individual muscle fibers (Plate 17), through the larger layers of perimysium that delineate bundles of muscle fibers (Plate 19), and culminating in thick, strong layers of epimysium on the surfaces of individual muscles (Schmitt et al., 1979).

Collagen Fibers

Under the light microscope, collagen fibers in the connective tissue framework of meat range in diameter from 1 to 12 μm. They seldom branch, and, if branches are found, they usually diverge at an acute angle. Collagen fibers from fresh meat are white, but they may be stained histologically with acid dyes such as eosin (which stains collagen fibers pink). Unstained collagen fibers may be seen by polarized light since they are birefringent, as explained in Chapter 12. Collagen fibers have a wavy or crimped appearance, which disappears when they are placed under tension.

Collagen fibers are composed of small fibrils that are formed from long tropocollagen molecules with a staggered arrangement and lateral covalent bonding. In preparation for electron microscopy, negative staining with heavy metals spreads into the spaces between the ends of tropocollagen molecules so that collagen fibrils appear to be transversely striated. The periodicity of striations is 67 nm but often shrinks to 64 nm as samples are processed for examination. Although collagen fibers are located outside the cell, the initial stages of collagen fibril formation probably are within the cell, with fibril morphology being regulated by a special site on the fibroblast membrane.

Tropocollagen

Tropocollagen has a high molecular weight (300,000 Daltons) and is formed from three polypeptide strands twisted into a triple helix. Each strand is a left-handed helix twisted on itself, but the three strands are twisted into a larger right-handed triple helix. The triple helix is responsible for the stability of the molecule and for the property of self-assembly into fibrous components. Telopeptides projecting beyond the triple helix are responsible for cross-linking between adjacent molecules, with cross-links to the shaft of the adjacent molecule.

In the polypeptide strands, glycine occurs at every third position, and proline and hydroxyproline account for 23% of the total residues. Hydroxyproline is relatively rare in other proteins of the body, so that an assay for this imino acid provides a measure of the collagen or connective tissue content of meat (O'Neill et al., 1979; Etherington and Sims, 1981). Tropocollagen also contains a relatively high proportion of glutamic acid and alanine, as well as some hydroxylysine.

Types of Collagen

The tropocollagen molecule is composed of three alpha chains, but nineteen unique alpha chains have been identified, giving rise to eleven different types of collagen. These may be categorized into three general classes: (1) molecules with a long (≈ 300 nm) uninterrupted helical domain, (2) molecules with a long (≥ 300 nm) interrupted helical domain, and (3) short molecules with either a continuous or an inter-rupted helical domain (Miller, 1985).

The important types of collagen for on-line evaluation of meat are Type I collagen, forming striated fibers between 80 and 160 nm in diameter in blood vessel walls, tendon, skin, and muscle; Type II, forming fibers less than 80 nm in diameter in hyaline cartilage; and Type III reticular fibers, found in muscle endomysium and around adipose cells. Other types of collagen are less important in on-line evaluation, although they may contribute to background fluorescence. Type IV collagen occurs in basement membranes. Although basement mem-branes once were regarded as amorphous, now many are thought to be composed of a network of irregular cords. The cords contain an axial filament of Type IV collagen, plus ribbons of heparin sulfate proteoglycan and fluffy material (laminin, entactin, and fibronectin). Type IV collagen occurs in the endomysium around individual muscle fibers. Instead of being arranged in a staggered array, the molecules are linked at their ends to form a loose diagonal lattice. Type V collagen also has been found in muscle fiber basement membranes. Type VI collagen, a tetramer of Type V, forms a filamentous network and has been identified in muscle and skin. The molecule consists of a short triple helix about 105 nm in length with a large globular domain at each end.

Tendons often extend into the belly of a muscle or along its surface before they merge with its connective tissue framework, and Types I and III collagen both may be extracted from meat. Even within tendons, there may be some Type III collagen forming the endotendineum or fine sheath around bundles of collagen fibrils. Small-diameter Type III collagen fibers are called reticular fibers because, when stained with silver for light microscopy, they often appear as a network or reticulum of fine fibers. Large-diameter Type I collagen fibers are not blackened by silver but are stained yellow.

Collagen Accumulation

Although the relative proportions of Types I and III collagen in meat may be related to tenderness (Burson and Hunt, 1986), the overall amount of collagen and its degree of cross-linking also are important. The absolute amount of collagen may increase with animal age, and this may have an effect on meat toughness, but rapid growth of muscle fibers dilutes the relative amounts of collagen in meat.

Within a carcass, there may be considerable differences in collagen content between different muscles, and this is reflected in the retail price of many meat cuts. Collagen content also may differ between sexes. However, the amount of collagen in meat, when expressed as a proportion of wet sample weight, is also affected by fat content. In veal steaks, for example, the collagen content might exceed 0.5% but could be much less in the same region from a steer carcass in which muscle and fat accumulation had diluted the collagen content.

Collagen Cross-linking

The three polypeptide strands of tropocollagen are linked by stable intramolecular bonds originating in the nonhelical ends of the molecule. Stable disulfide bonds between cystine molecules in the triple helix also occur. Collagen fibers owe much of their tensile strength, however, to the covalent bonds between adjacent tropocollagen molecules. During the growth and development of meat animals, covalent cross-links increase in number, and collagen fibers become progressively stronger. Meat from older animals, therefore, tends to be tougher than meat from the same region of younger animals. This relationship is complicated in young animals by the rapid synthesis of large amounts of new collagen. New collagen has fewer cross-links so that, if there is a high proportion of new collagen, the mean degree of cross-linking may be low, even though all existing molecules are developing new cross-links. As the formation of new collagen slows down, the mean degree of cross-linking increases.

Another complication is that many of the intermolecular cross-links in young animals are reducible (the collagen is strong but is fairly soluble). In older animals, reducible cross-links are probably converted to nonreducible cross-links (the collagen is strong but is far less soluble and more resistant to moist heat). Changes in collagen solubility appear to have a greater impact on meat tenderness in beef from older animals than younger animals (Hall and Hunt, 1982).

Despite some initial uncertainties in the 1980s, pyridinoline now is widely recognized as a nonreducible trifunctional cross-link of collagen, and, in the medical field, its presence in the urine is used as a marker of pathological collagen degradation (Fincato et al., 1993). Pyridinoline is involved in the increased heat stability of connective tissues from older animals (Nakano et al., 1985). From the work of Zimmerman et al. (1993), one would expect slow muscles in unexercised animals to have a greater rate of pyridinoline formation than fast muscles. Relative to

USDA beef maturity levels, the pyridinoline content and thermal stability of intramuscular collagen both increase together (Smith and Judge, 1991).

Differences in the degree of cross-linking may occur between different muscles of the same carcass and between the same muscle in different species. For example, collagen from the longissimus dorsi is less cross-linked than collagen from the semimembranosus, and collagen from the longissimus dorsi of a pork carcass is less cross-linked than collagen from the bovine longissimus dorsi. Nutritional factors such as high-carbohydrate diet, fructose instead of glucose in the diet, low protein, and preslaughter feed restriction may reduce the proportion of stable cross-links. Nonenzymatic glycosylation (a reaction between lysine and reducing sugars) may be involved in the interaction between diet and collagen strength. In general, the turnover rate of collagen is accelerated in cattle fed a high energy diet. The rate of collagen turnover in skeletal muscle may be about 10% per day, and the turnover time for collagen may be inversely proportional to collagen fibril diameter.

Elastic Fibers and Elastin

Individual collagen fibers only lengthen by about 5% when stretched, and little elasticity is possible where collagen is formed into cable-like tendons. However, much of the collagen that is present in meat forms a meshwork, so that stretching causes configurational changes. Fibers with truly elastic properties, however, are necessary in structures such as the ligamentum nuchae and abdominal wall. And all arteries, from the aorta down to the finest microscopic arterioles, rely on elastin fibers to accommodate the surge of blood from contraction of the heart. Elastin fibers may be stretched to several times their original length, but they rapidly resume their original length once released.

Elastin is the protein from which elastic fibers are composed. Elastin resists severe chemical conditions, such as the extremes of alkalinity, acidity, and heat that destroy collagen. Fortunately, there are relatively few elastic fibers in muscle; otherwise, cooking would do little to reduce meat toughness. The elastin fibers in muscles used frequently for locomotion are larger and more numerous than those of less frequently used muscles. Elastin fibers in the epimysium and perimysium of beef muscles range from 1 to 10 μm in diameter. Elastin is synthesized by arterial smooth muscle cells, but the origin of elastin in nonvascular locations is not properly understood. In the lung, for example, large amounts of elastin are synthesized by various types of lung cells, but the cellular source of the elastin fibers in meat is unclear at present. Some elastic fibers in muscle are involved in the attachment of neuromuscular spindles.

Elastic fibers are pale yellow and birefringent. In the bovine ligamentum nuchae, the pattern of birefringence indicates that there are two micellar structures, one arranged circularly on the outside and the other arranged axially in the centers of the fibers. Elastic fibers in meat have a small diameter (approximately 0.2 to 5 μm) although they are much larger in the ligamentum nuchae. Elastic fibers in the connective tissue framework of meat are usually branched.

Electron microscopy reveals that elastic fibers are composed of bundles of small fibrils approximately 11 nm in diameter embedded in an amorphous material. In the bovine ligamentum nuchae, fibrils may be constructed from smaller units or filaments approximately 2.5 nm in diameter. Elastin filaments are bound by noncovalent interactions to form a three-dimensional network, and elastic fibers are assembled in grooves on the fibroblast surface where initially rope-like aggregations of fibrils become infiltrated with amorphous elastin. Unlike the situation in elastic ligaments where elastin forms fibers, the elastin of the arterial system occurs in sheets that condense extracellularly in the absence of fibrils.

Although elastin resembles tropocollagen in having a large amount of glycine, it is distinguished by the presence of two relatively rare amino acids, desmosine and isodesmosine. Like collagen, elastin contains hydroxyproline, although it may not have the same function of stabilizing the molecule. Tropoelastin, the soluble precursor molecule of elastin (70,000 to 75,000 Daltons), is secreted by fibroblasts after it has been synthesized by ribosomes of the rough endoplasmic reticulum and processed by the Golgi apparatus. In the presence of copper, lysyl oxidase links together four lysine molecules to form a desmosine molecule. Isodesmosine is the isomer of desmosine.

Cooking

Collagen fibers shrink when they are cooked, and, ultimately, they may be converted to gelatin. At approximately 65°C, the triple helix is disrupted, and the alpha chains are randomized in arrangement. The importance of this change is that it tenderizes meat with a high connective tissue content. Tropocollagen molecules from older animals are more resistant to heat disruption than those from younger animals. At an intermediate level of cooking around 65°C, endomysium and perimysium may differ in the extent to which they are altered by heat because of differences in cross-link thermal stability. Heat-induced solubilization of Type I collagen is more important in improving meat tenderness by cooking than is the effect of heat on Type III collagen.

CAUSES OF BEEF TOUGHNESS

Having introduced some of the basics of connective tissue biology in meat, it is important to see how connective tissues might fit into the multifactorial system responsible for beef texture. There are many factors that contribute to toughness in beef, including:

(*1*) Animal age (Bouton et al., 1978)

(*2*) Sarcomere length (Herring et al., 1967; Dutson et al., 1976; Marsh and Carse, 1974)

(*3*) Postmortem pH (Bouton et al., 1973a, 1973b; Jeremiah et al., 1991)

(*4*) The condition of the cytoskeleton (Davey and Graafhuis, 1976)

(*5*) The temperature coefficient (Q_{10} effect) of aging (Davey and Gilbert, 1976)

(*6*) The temperature at the onset of rigor mortis (Locker and Daines, 1975a; Martin et al., 1977)

(*7*) Factors that affect postmortem proteolysis (Etherington, 1984; Wheeler et al., 1990; Koohmaraie et al., 1991; Koohmaraie, 1992; Goll et al., 1992; Ouali, 1992)

(*8*) Nonenzymatic postmortem tenderization (Takahashi, 1992)

(*9*) Muscle shortening during cooking (Locker and Daines, 1975b; Davey and Gilbert, 1975)

Also, there are possibilities for improving tenderness by calcium chloride injection and other techniques (Whipple and Koohmaraie, 1992).

The connective tissue contribution to beef toughness originates during production, rather than during the uncertainties of transport, slaughtering, and refrigeration. It may involve elastin (Bendall, 1967) and collagen concentration (Dutson et al., 1976), the distribution of different types of collagen (Light et al., 1985; Burson and Hunt, 1986), collagen cross-linking (Goll et al., 1964; Smith and Judge, 1991), collagen orientation (Rowe, 1974; Locker and Daines, 1976; Dransfield and Rhodes, 1976), thermal contraction during cooking (Bailey and Sims, 1977), and thermal survival during cooking (Carroll et al., 1978).

CONNECTIVE TISSUE FLUORESCENCE

Biochemistry

It is common knowledge among histologists that both collagen and elastin fibers are fluorescent, emitting blue-white light when excited with UV (Pearse, 1972). Differential quenching of fluorescence with phosphomolybdic acid enables the separation of elastin from collagen (Puchtler et al., 1973). With excitation at 335 nm and 370 nm and measurement at 385 nm and 440 nm, respectively, the fluorescence of collagen in rats increases exponentially with age ($r = 0.83$ and 0.90, respectively) and is a reliable marker for biological age (Odetti et al., 1992). The cause of this age-related increase in fluorescence is thought to be associated with cross-linking and polymerization from glycosylation by reducing sugars (Schnider and Kohn, 1981; Monnier et al., 1984; Brownlee et al., 1986; Vishwanath et al., 1986; Oimomi et al., 1986; Monnier, 1989). Fluorometry of tyrosine, proline, and hydroxyproline in homogeneous collagen solutions is used for biochemical research on the structure of collagen (Bellon et al., 1985; Na, 1988) but at much lower wavelengths than are feasible at present with fiber optics in meat. However, this is a likely area for future research because UV optical fibers are steadily improving. Elastin contains cross-linking amino acids that are fluorescent (Suyama and Nakamura, 1992).

Microspectrofluorometry

During the 1980s, as biochemists developed collagen fluorescence as a method for measuring biological age, it seemed reasonable to start looking at the fluorescence of collagen and elastin in meat. Microscope spectrofluorometry takes a while to master, especially if building and programming one's own apparatus, but eventually some relatively simple techniques were established, which were then adapted for on-line application using fiber optics. As a development strategy, this offered two important advantages: first, working under the microscope enables the exact type of tissue generating a fluorescence spectrum to be determined (Swatland, 1987a), and, second, many of the techniques of microscope photometry are readily adaptable to working with optical fibers (Swatland, 1987b).

Although microscopy is useful for identifying the microstructural sources of fluorescence in a meat product, the fluorescence emission spectrum of a standard tissue such as bovine tendon may vary considerably in shape when measured by microscopy at the semi-micro level using a fiber-optic meat probe or at the macroscopic level using a quartz rod (Chapter 12). Unfortunately, fluorescence measurements are difficult to standardize, and they tend to reflect the apparatus with which they are recorded as much as the nature of the sample (Bashford, 1987). The principles and methodology of microscope photometry for a manually operated system are well known (Piller, 1977), and an accepted manual method for the measurement of relative spectral fluorescence intensities (Zeiss, 1980) may be used as the main algorithm for both microscopy and meat probes. Eventually, it proved possible to obtain similar fluorometry spectra with both microscopy and meat probes.

There are two different methods to obtain information about the connective tissue content of meat using fluorometry. One method is to obtain stereological information on the distribution of collagen in meat, taking a broadband measurement of the overall fluorescence emission through a small window of a moving probe as an indicator of collagen distribution. The other method is to measure the fluorescence emission spectrum of the meat through a large stationary window, using the shape of the spectrum to obtain information on both the amount and biochemical type of the collagen in the field. The latter method depends on the fact that biochemical Types I and III collagen have different fluorescence emission spectra (Swatland, 1987a, 1987b). Maximum excitation for most types of collagen in meat is at 375 nm (Figure 9.1), although 370 nm may give the best separation of Types I and III collagen using a ratio of emission at 440 to 510 nm (Swatland, 1987c). Thus, for practical purposes, excitation may be obtained with the 365-nm peak of a mercury lamp. But Type I fibers emit a prequenching spectrum for longer than Type III fibers, probably because Type I fibers have a larger diameter than Type III fibers. The relative intensities of pre- and postquenching emission spectra may be quantified by taking a ratio (such as 440:510 nm). In this case, the fluorescence of elastin is added to the prequenching peak around 440 nm, which is associated with Type I collagen fibers (Swatland, 1987a). This peak is an indicator for connective tissues that may cause meat toughness. On the other hand, the fluorescence of the

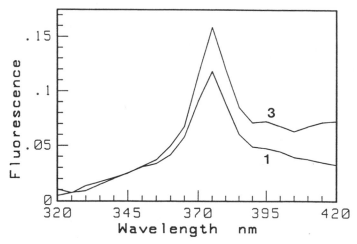

Figure 9.1 *Fluorescence excitation spectra for purified Type I (line 1) and Type III (line 3) collagen.*

endomysium is combined with the fluorescence of Type III collagen fibers around adipose cells (Swatland, 1987e). Both of these contribute to the postquenching peak at 510 nm, which is regarded as the background fluorescence of meat with low connective tissue toughness. This is a fortunate coincidence that enables fluorescence to be used to predict connective tissue toughness in meat, but this method may not work properly if technical factors change the shape of the emission spectrum, as described below.

Figure 9.2 shows four fluorescence emission spectra of connective tissue fibers, all measured in the same way. The fibers were teased apart and mounted in distilled water under a cover slip, and a small measuring aperture was used to ensure that the measured field was evenly filled. An epifluorescence condenser was used with a UG1 excitation filter as the excitation monochromator and a dichroic mirror (FT395) as the beam splitter, essentially using the same apparatus as that shown in Figure 3.13, but with a microscope instead of optical fibers. In Figure 9.2, spectra 1 and 2 are for chemically purified fibers of Types I and III collagen respectively, which are the dominant connective tissue fibers of extramuscular tendon and intramuscular connective tissue shown by spectra 3 and 4. Although the tissue types are not pure (extramuscular tendon may contain Type III collagen and intramuscular connective tissue may contain Type I collagen), there is an appreciable degree of separation between the spectra. Thus, there are certain wavelength ratios

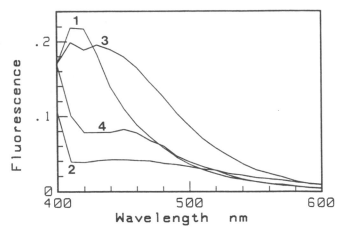

Figure 9.2 *Fluorescence emission spectra of isolated fibers measured by microscope spectrofluorometry; purified Type I (line 1), purified Type III (line 2), tendon collagen fibers (3), and collagen fibers from intramuscular connective tissue (4).*

that may be used to determine the identity of the tissue. However, the two strong spectra (1 and 3) show an increase from 400 to 410 nm, while the two weak spectra (2 and 4) show a decrease. Although all four spectra were measured in the same way and spectra 1 and 3 appear reasonable, do spectra 2 and 4 really have a maximum at 400 nm? Answering this problem in microscope spectrofluorometry (Swatland, 1990) is important because it helps explain the basis of meat probe measurements.

The fluorescence blank is a measurement of the background fluorescence in the absence of a sample and is required because optical components may be contaminated with dust (keratin dust from human and animal skin is strongly fluorescent). Stray fluorescence also may originate from optical components and media (such as immersion oil in microscopy), or there may be cross-talk between outgoing and ingoing optical fibers. Thus, it is important to measure an appropriate fluorescence blank at each wavelength, to store these data as a vector, and to subtract them from the vector obtained on measuring a sample.

Unfortunately, the excitation maximum for connective tissues is around 370 nm, which is fairly close to the emission peak at 410 to 430 nm. With an excitation filter and a dichroic mirror, problems are caused by the reflectance from the sample of the upper limit of the excitation bandpass. For both UG1 and UG5 excitation filters used in fluorometry of connective tissues, the upper limit exceeds 400 nm (Zeiss bulletin 41-305-e). In samples with a low fluorescence intensity (such as spectra

2 and 4 in Figure 9.2), reflectance from the sample of the upper limit of the excitation bandpass might occur at 400 nm. By deduction, this would imply that the corresponding segments of spectra of samples with a high fluorescence intensity also are unreliable and that their shape is determined as much by the apparatus as by the fluorescence of the sample. Thus, the data for low wavelengths in fluorescence emission spectra for connective tissue fibers in meat are unreliable because they are shaped by the type of fluorescence blank used. A perfect blank would have the same reflectance as the sample, thus enabling the separation of fluorescence from the reflectance of the high-wavelength edge of the excitation spectrum.

When making broadband fluorescence measurements stereologically using a meat probe, the minimum value of the signal (a window of muscle devoid of connective tissue fluorescence) is the sum of all the light that short-circuits the full optical path through the meat. Most originates from light at the crossover wavelengths of the excitation filter, dichroic beam splitter, and the barrier filter, but it is supplemented by stray fluorescence from dust and optical fibers. Operationally, this pseudofluorescence functions in the same way as the small light guide linking the reference photodiode to the zenon flash in Figure 3.11 and may be exploited to provide information on the intensity of the illuminator.

Connective Tissue in Comminuted Meat

Connective tissue levels in ground beef are sometimes a problem if too many meat scraps with a high content of tendon are worked into a product. The result may be a gritty texture for hamburger or excessive gelatin formation in a cooked product. Elastin derived from elastic ligaments has virtually the same fluorescence emission spectrum as Type I collagen from tendon and ligaments, which enables fluorescence emission ratios to be used to predict total problematic connective tissue levels.

Calibration spectra are shown in Figure 9.3, and, as described above, the secondary peak of pseudofluorescence at 420 nm may be ignored because it is created by a type of fluorescence blank. There are several wavelengths that may be used to give a ratio that describes the connective tissue content (Swatland, 1987d). At 510 nm, both spectra are very close, so this wavelength corrects for any drifting in the intensity of excitation. A wavelength from the region where the two spectra reach maximum divergence contains information on the connective tissue level

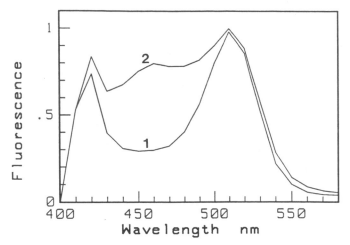

Figure 9.3 *Fluorescence emission spectra of ground beef from sources with very low (1) and very high (2) connective tissue contents.*

in the meat (440 nm was chosen to be consistent with earlier research separating Types I and III collagen). The ratio of these two wavelengths is a reliable predictor ($r = 0.93$) of intermediate mixtures of ground beef from the two calibration sources. Differences in fat content are not a problem because the reticular fibers around adipose cells have similar fluorescence to the reticular fibers of the endomysium: essentially, fat is indistinguishable fluorometrically from muscle devoid of major connective tissues. As discussed in Chapter 13, feedback control may be used to keep the connective tissue level constant in a continuous processing operation. Thus, the incorporation of meat scraps with a high connective tissue content may enable a least-cost formation, while staying below the detection threshold for the final product.

CONNECTIVE TISSUE PROBE

Fluorescence sensors are used for many different applications in industrial biotechnology (Kuhn et al., 1991; Li and Humphrey, 1991; Siano and Mutharasan, 1991; Wang and Simmons, 1991a, 1991b), and a meat probe fitted with optical components similar to those shown in Figure 3.12 may be used to obtain a stereological sample of the connective tissues in meat, working on meat cuts or whole carcasses in a meat cooler (Swatland, 1991b). The optical measurement is a broadband measurement of the whole fluorescence emission spectrum plotted against depth in the meat. Thus, when the optical window of the probe penetrates or passes a seam of connective tissue, it produces a peak in the fluorescence signal, as described in Chapter 3. Figure 9.4 shows the comparison of extremely tough round steak compared with extremely tender rib-eye, both taken from a large taste-panel experiment on Canada Grade A beef with a range in animal age from twelve to twenty-four months (Swatland et al., 1994). For samples at intermediate toughness levels within this range, signals had peaks with an intermediate size and distribution so that comparisons between samples were made on the statistical analysis of signal peaks.

Effect of Temperature and Freezing

Fluorescence emission is insensitive to temperature, while fluorescence quenching is proportional to temperature (Harris and Bashford, 1987). Thus, fluorescence generally is stronger at low temperatures than at high temperatures. In an on-line application, working on meat that has been chilled to a constant temperature but not frozen, temperature is not critical, provided that it is relatively low and constant. However, a reverse temperature effect may appear in meat that has been frozen and

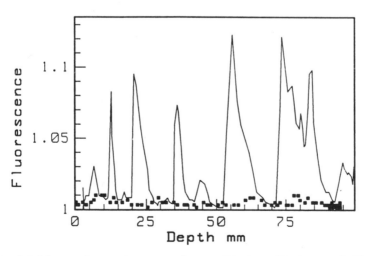

Figure 9.4 Meat probe measurements of connective tissue fluorescence in Canada Grade A beef, showing results from the toughest semitendinosus muscle (solid line) compared with the most tender longissimus dorsi muscle in a taste-panel trial.

then thawed to a typical meat-handling temperature of 4 to 5°C, and fluorescence may increase as the meat temperature increases during thawing (Figure 9.5).

The problem was encountered when probe measurements were made on samples that previously had been frozen and then thawed. Measurements made shortly after the meat became soft enough to penetrate with the probe had no relationship to the known collagen content of the samples, but, when remeasured later in the day, strong correlations were detected. This emerged as a difficult problem to investigate because it is difficult to measure the same sample twice: exposure to UV light for the first measurement quenches some of the fluorescence so that subsequent fluorescence measurements get progressively weaker.

A typical experiment is shown in Figure 9.6. Fluorescence emissions (435 to 445 nm) were measured at regular intervals as beef tendon (starting at room temperature) was frozen, then warmed to body temperature, then cooled back to room temperature. A shutter was used to protect the sample from UV between the measurements, but quenching caused by repeated measurement still occurred. Thus, at the end of the experiment, the sample had much less fluorescence than at the start of the experiment. When the sample first was cooled, lower temperatures slightly reduced the quenching effect of repeated measurement. The thawing anomaly is seen at point A in Figure 9.6, where fluores-

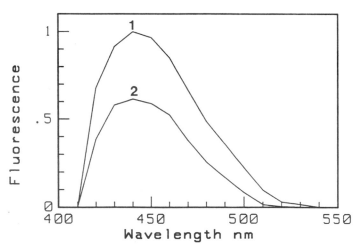

Figure 9.5 *Anomalous effects of temperature on connective tissue fluorescence. For Type I collagen of tendon, the stronger emission spectrum (1) was at 5°C while the weaker spectrum (2) was at the point of thawing ≤0°C, instead of vice versa.*

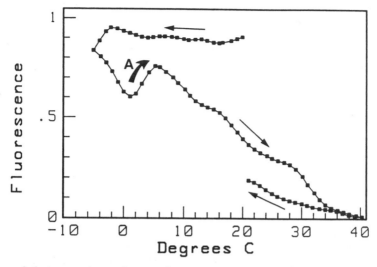

Figure 9.6 *An experiment showing fluorescence emissions (435 to 445 nm) of beef tendon, quenched by repeated measurement but also affected by temperature. The arrows show the direction of measurement, and the thawing anomaly is at point A.*

cence temporarily increased despite increasing temperature and repeated measurement.

Investigation of the problem (Swatland, 1993a) enabled a working hypothesis to be proposed. Light scattering seems to be reduced as samples thaw from <0° to 5°C so that: (1) UV excitation penetrates farther into the tissue, (2) fluorescence emissions may be captured from deeper in the tissue, (3) fluorescence from structures near the optical window is not lost by scattering, and (4) there is an increase in the detection of distant fluorescent structures subtending a small angle at the optical window. The causative mechanism might be the melting of ice crystals, temporarily creating clear optical pathways at the microstructural level. Whatever the merits of this hypothesis, however, there is no doubting the practical importance of the effect: unless meat is completely thawed so that all ice crystal pockets have gone, its connective tissue fluorescence cannot be measured properly.

Standardization of Data

One of the major problems of fluorometry in general is the difficulty of standardization. Working with a relatively simple system such as a dilute fluophor in a cuvette, it may be possible to express fluorescence as a quantum yield, but this is virtually impossible with an optical fiber pushed into a sample of meat with an unknown light path through the tissue. All we know at present is that the UV output of probes that work effectively is quite high, on the order of 100 μW. Thus, as explained earlier in the section on microspectrofluorometry, the probing of meat to identify major connective tissue distribution patterns is based on the quenching of background fluorescence from Type III collagen, and the detection of masses of Type I collagen fibers still with a prequenching emission spectrum. Thus, an immediate problem in making connective tissue fluorescence measurements on-line is how to standardize the measurements to allow for aging of the UV source. One solution would be to use an internal reference, as in Figure 3.11, but this has not yet been investigated. The option that has been investigated, however, is to use pseudofluorescence.

An experiment was undertaken to push the probe far beyond its normal working limits by making major changes in UV intensity, hoping to find a software solution, which then could be used with confidence to correct for minor variations in UV intensity in everyday usage (Swatland, 1993b). Samples of longissimus dorsi and semitendinosus (all Canada Grade A) were measured at 4°C forty-eight hours postmortem. The two different muscles were measured rapidly in an alternating sequence with a stabilized UV source to enable a drift-free comparison of the two muscles. As expected, semitendinosus produced signals with stronger and more numerous fluorescence peaks than longissimus dorsi. Then the muscles were remeasured separately in two batches using contradictory levels of UV intensity so that semitendinosus produced signals about 100 times weaker than longissimus dorsi.

Part of the software solution also deals with the problem of rejection level. From Figure 9.4, it is evident that the larger peaks in the signal contain the required information about the major connective tissues in the sample: the problem is where to set the rejection threshold to accept large peaks and reject small ones. Results from raw data were compared with results from data that were scaled by dividing all the fluorescence measurements in a transect by the minimum fluorescence (mostly pseudofluorescence) in each transect. The minimum fluorescence (Y_{min}) also was used to specify a cutoff level for the suppression of small background peaks, using an arbitrary threshold (0.1).

$$\text{cutoff} = Y_{min} + (Y_{min} \times 0.1)$$

Three other thresholds were tested (0.1, 0.2, and 0.3), and they all gave similar results. Results were evaluated statistically with a paired t-test for semitendinosus versus longissimus dorsi so that differences between animals were unimportant.

The experimental bias caused by using contradictory UV intensity was only partly cancelled by scaling the data, and the numbers and widths of major fluorescence peaks were not restored to their true values. Fluorescence intensities or peak heights, however, were corrected by scaling. The effects of contradictory UV intensity are worth examining in detail because they show the limitations of the technology.

Contradictory UV intensity increased the fluorescence intensities of peaks in the more strongly excited longissimus dorsi, which was expected. But other results were not anticipated. Because the fluorescence of collagen fibers fades fairly rapidly with continued exposure to UV and because large- and small-diameter collagen fibers have different survival times for their prequenching emission spectra, the number of peaks per centimeter increased in longissimus dorsi as a result of higher intensity UV excitation. This may have been caused by an increased depth of UV penetration reaching more seams of collagen. Paradoxically, however, the number of peaks also increased in semitendinosus when UV excitation was decreased, which may have been caused by slower quenching of the fluorescence. The net result was that, instead of semitendinosus having more peaks per centimeter than longissimus dorsi (correct), contradictory UV intensity caused the longissimus dorsi to have more peaks per centimeter than semitendinosus (incorrect).

In summary, it is very difficult to adjust data from a fluorescence probe for major variations in the intensity of UV excitation because this interacts with both the degree of quenching and the depth of UV penetration. The best we can do at present is to scale the data, dividing fluorescence measurements by the minimum value (pseudofluorescence) in each transect. This corrects the intensities of peaks satisfactorily but is less effective for the numbers and widths of peaks; however, if variation in UV intensity is relatively slight, then scaling is the best way to compare data collected in different batches.

Correlation with Collagen Content

Selecting a threshold to calculate a cutoff level to reject small peaks in the signal is a lot easier than dealing with major variations in UV intensity. Using a batch of typical commercial meat samples with biochemically determined collagen levels, the simple correlation coefficients of mean peak intensity, width, and incidence may be plotted as a function of threshold level (Swatland et al., 1993). As shown in Figure 9.7, all three features of the signal reach a maximum relationship to collagen content ($r \geq 0.85$) over a threshold of 0.14. This confirms what one might reasonably suspect, that it is the occurrence of the larger seams of connective tissue on which the correlation of fluorescence with collagen is based. Since this is the case, it is possible to use not just the mean value, but the intensities of the largest peaks in a transect as well. Thus, for the same experiment as in Figure 9.7, using stepwise multiple regression, four features of the signal were selected that are strongly correlated ($R = 0.93$) with the collagen concentration; namely, the intensities of the largest two peaks in the transect, the mean intensities of peaks above the cutoff, and the area under the curve above the cutoff.

As explained in Chapter 3, the shape and magnitude of a fluorescence peak are determined by both the size and fluorescence intensity of the connective tissue structure generating the peak. Maximum pyridinoline fluorescence is at lower wavelengths than are currently accessible by fiber optics in meat (excitation maximum at 295 nm, with emission from 400 to 410 nm), but the upper tail of the spectrum may be detectable in

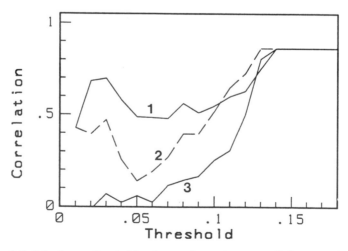

Figure 9.7 *Selecting a threshold value to determine a cutoff for accepting large fluorescence peaks in UK beef roast, using the simple correlations of mean peak intensity (1), mean peak width (2) and mean incidence of peaks per centimeter (3) with biochemically determined collagen content.*

meat. Thus, the height of the largest peak in the experiment of Figure 9.7 was correlated with the pyridinoline (corrected for differences in collagen concentration), $r = 0.77$ ($R = 0.86$ by stepwise regression).

Correlation with Taste-Panel Data

Canadian supermarket customers rate tenderness very highly as a key attribute of beef quality that needs to be improved and made more consistent (McDonell, 1988). Thus, there is a strong commercial incentive to improve beef tenderness. However, until the experiment reported below, there was little or no evidence that sporadic toughness in commercial Grade A beef could be caused by connective tissue, and most Canadian meat scientists had concluded, like their colleagues in the United States, that beef toughness must have some other cause, such as a rapid growth rate reducing degradative enzyme activity.

Three age groups of Canada Grade A carcasses were assembled based on production records and dentition: at approximately twelve months ($n = 59$, carcass weight 293 ± 37 kg), seventeen months ($n = 54$, 344 ± 43 kg), and twenty-four months ($n = 54$, 352 ± 46 kg). In rib-eye and round steaks, age-related effects in connective tissue distribution were detectable with the meat probe. From twelve to seventeen months, the incidence of fluorescence peaks per centimeter increased, probably as layers of perimysium developed and became detectable. From seventeen to twenty-four months, the incidence of fluorescence peaks per centimeter decreased, probably as layers of perimysium were pushed apart by muscle fiber growth. Fluorescence peaks became wider from seventeen to twenty-four months, probably as layers of perimysium grew thicker.

A search was made for relationships between probe signals and taste-panel evaluations of meat toughness (chewiness). At twelve, seventeen, and twenty-four months, absolute values for the strongest simple correlations of chewiness with the probe signal were, respectively, $r = .32$ ($P < .01$), $.32$ ($P < .01$), and $.47$ ($P < .0005$) for semitendinosus and $.31$ ($P < .01$), $.29$ ($P < .05$), and $.18$ (NS) for longissimus dorsi. The corresponding multiple correlations derived from stepwise regressions of indices were $R = .64$, $.61$, and $.86$ for semitendinosus and $.63$, $.47$, and $.65$ for longissimus dorsi.

Obviously, connective tissue is not the sole cause of variation in tenderness in Canada Grade A beef, but it appears to have some effect, exactly where one might anticipate, in the older animals allowed into the grade and in cuts of meat with an intermediate level of connective tissue toughness. Nobody is suggesting that this technology in its present form should be used for statutory grading, but it would be foolish to ignore the commercial possibilities that it offers. Thus, for a plant that hand-picks beef for a premium treatment, this technology could help considerably in avoiding any meat with excessively high connective tissue levels. Thus, taking the twenty animals that the taste panel found had the toughest meat (all ages pooled) and comparing them with the remainder, the tough carcasses had a higher mean peak intensity than the others (0.027 ± 0.013 versus 0.020 ± 0.010, respectively, $P < 0.005$). Bearing in mind that a measurement takes only a couple of seconds in a meat cooler and is virtually nondestructive, this technology has some potential as an on-line method to predict meat toughness originating from connective tissue, at least in Canada, if not in the United States.

In countries where beef is a by-product of the dairy industry, the commercial justification for using fluorescence probe technology to sort beef is a lot stronger, because everyone agrees that connective tissues may make a substantial contribution to the toughness of beef from old cows. In a recent trial of Danish beef (Swatland et al., 1995), a taste panel tested samples of longissimus dorsi and semitendinosus muscles from twenty beef carcasses covering a wide chronological age. In longissimus dorsi, the number of fluorescence peaks per centimeter was correlated with toughness ($r = 0.73$, $P < 0.0005$), whereas the half-width of peaks was the best single predictor for semitendinosus ($r = 0.66$, $P < 0.005$). Using stepwise regression of a number of features of the signal stronger prediction equations are possible ($R = 0.95$ for meat toughness).

Correlations of taste with the fluorescence signal also were detected ($r = 83$, $P < 0.0005$ for peaks per centimeter), although they might have been secondary to a correlation of taste with toughness ($r = 61$, $P < 0.005$). However, semitendinosus juiciness was correlated with the height of the largest fluorescence peak in a transect ($r = 0.78$, $P < 0.0005$) even though no significant correlations of semitendinosus juiciness with toughness were detectable ($r = 0.06$). Thus, from this preliminary study, it appears that the UV probe works as well on a wide age range of carcasses as it does on a narrow age range. However, whether or not this technology will survive the transfer from a pilot plant to the real world remains to be seen, but at least it is worth a try.

Bidirectional Operation

As explained in Chapter 4, the rheological properties of meat are important in understanding the operation of meat probes. Using the terminology for fat-depth probes introduced in Chapter 1, way-in and way-out transects were concatenated for Figure 9.8, so that the double peaks originate from single anatomical structures but with a distance offset caused by the probe deforming the meat. Distance offsets in paired peaks may be seen in nearly all bidirectional transects, and they are absent when the probe is tested against a bar chart, as described in Chapter 3. Of course, if a probe is not working properly, a distance offset might be encountered in a test against a bar chart.

In Figure 9.8, the pairs of major peaks are labeled: A,a and B,b. Peak A was encountered first on the way-in, followed by peak B deep in the meat. Both peaks were detected again on the way-out, first peak b deep in the meat and then peak a. Note that the depth is a relative measurement made between the probe shaft and a plate on the surface of the meat. On initial penetration, the meat is compressed in advance of the blade but stretched between the blade and the surface plate, so that peaks A and B appear to be farther into the meat than they really are. The reverse effect occurs as the probe is withdrawn, since the meat from which the probe

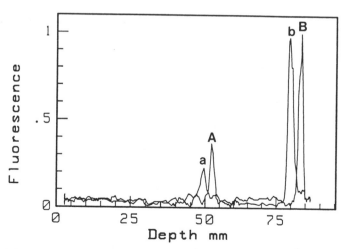

Figure 9.8 *Meat probe measurements of connective tissue fluorescence in beef semi-membranosus muscle with a high connective tissue content. Two intense sources of fluorescence were detected on the way-in (A and B) and on the way-out (b and a).*

had retreated is stretched, while the meat between the blade and the surface plate is compressed. Thus, for way-out measurements, the peaks appear to be nearer the surface of the meat than they really are.

Also of interest in Figure 9.8 are the differences in peak height, particularly peak A versus a. Computer-assisted testing of the probe shows that this effect did not originate from the probe itself (measured fluorescence intensity was correlated with actual intensity, $r > 0.95$), but from an interaction of the probe with the meat. This effect might be caused by fluorescence quenching, but there are other possibilities, such as microstructural damage.

Distortion of the meat by a probe is a feature of this technology that needs a lot more attention. Fat-depth probes generally make measurements on the way-out because of this problem, but it might not be a perfect solution. As considered in the dynamic analysis of depth vectors in Chapter 4, the overall response of a probe may involve the rheological properties of the meat in unseen ways. For example, in Figure 9.8, the seams of connective tissue that generated peaks A and B might also have caused the probe to bounce, so that the peaks would appear wider or even doubled. A longer dwell time of the peak relative to the photometer might also create a stronger measured intensity for the peak. By doubling the effect of major peaks, therefore, the rheological properties of the meat may enhance an optical detection system. This effect has been detected in probe testing. In certain situations, probe signals may be strongly correlated with collagen content, but the correlation may be weakened by removing the disordered data from transects. Thus, while fat-depth probes may work best on the way-out, fluorescence probes for connective tissue may work best on the way-in, as in the example shown in Table 9.1.

Uniformities versus Singularities

It may not be feasible to keep the length of a probe transect constant, because the probe may hit a bone or come out of the other side of the meat. Thus, it is important to distinguish between those features of the signal that we may call singularities versus those we might term uniformities. A singularity is an infrequent and conspicuous feature in a signal, such as the occurrence of one or a few major peaks, like peaks A and B in Figure 9.8. The probability of detecting a singularity is directly proportional to transect length, and the absence of a singularity in a short

Table 9.1 *Comparing Way-in and Way-out Correlations of Fluorescence Signals (5 kHz damping) with Taste-Panel Evaluations of Danish Beef with a Wide Range in Age, Using Stepwise Regression of Signal Features to Calculate a Multiple Regression Coefficient,* $P < 0.01$ *for* $R > 0.7$ *(Swatland et al., 1995).*

Threshold Level	Way-in			Way-out		
	Hardness	Taste	Toughness	Hardness	Taste	Toughness
Orientation: coaxial with muscle fibers						
1.02	0.84	0.58	0.83	0.55	0.68	0.52
1.04	0.84	0.58	0.91	0.55	0.62	0.52
1.06	0.87	0.73	0.85	0.55	0.62	0.52
1.08	0.91	0.78	0.94	0.55	0.85	0.52
1.10	0.91	0.83	0.95	0.80	0.36	0.83
Orientation: perpendicular to muscle fibers						
1.02	0.68	0.47	0.72	0.54	0.41	0.59
1.04	0.70	0.89	0.73	0.55	0.41	0.59
1.06	0.68	0.60	0.72	0.55	0.41	0.59
1.08	0.68	0.89	0.72	0.35	0.41	0.34
1.10	0.68	0.93	**0.85**	0.40	0.41	0.53

transect is ambiguous (either there was no major connective tissue septum or else those that were present were missed).

A uniformity is the opposite of a singularity and is something that occurs regularly all the way along a transect so that it can be estimated reasonably well from any part of the transect, like the small peaks around 25 mm depth in Figure 9.8. For example, with a signal that has large peaks that rise above a background of small peaks (if there are no valleys or large negative peaks), a feature such as a regularly occurring minimum value is a uniformity.

Somewhere between singularities and uniformities, there may occur intermediate level patterns in the signal, such as frequent peaks of a moderate size. Parameters obtained from the analysis of patterns are far less dependent on transect length than are singularities, but they cannot be estimated reliably from very short transects.

The range from singularities to uniformities is rather arbitrary, but so is the range from macroscopic epimysium to microscopic endomysium to which the probe responds. Measurements made through extramuscular components of the carcass have very large singularities originating from major connective tissue structures, such as the aponeurosis over the longissimus dorsi. Singularities are useful in the prediction of overall collagen and taste-panel evaluations of tenderness, whereas the signal relating to pyridinoline residues per molecule of collagen tends to be closer to the uniformities (smaller elements of connective tissue) than is the information relating to bulk collagen (Swatland et al., 1993).

Effect of pH

When a fluorescence probe signal is correlated with collagen, part of the unexplained variance in the signal may originate from elastin, which is strongly fluorescent and may vary between carcasses (Bendall, 1967), but pH is also another possibility. Uniformities such as peaks per centimeter may be correlated weakly with pH ($r \approx 0.47$), but correlations with patterns and singularities are far weaker. Adjusting the data for pH may improve the prediction of total collagen (Swatland et al., 1993).

The scattering of light by meat is inversely proportional to pH because of an intrinsic effect of pH on the birefringence of myofibrils (Chapter 5), so that transmittance through muscle fibers is decreased at a low pH, with light at 400 nm showing a greater effect than light at 700 nm. Thus, the penetration of UV light to excite fluorescence may be lower in meat with a low pH than in meat with a high pH. Similarly, in this case, less of the fluorescence is likely to be transmitted back to the probe window in meat with a low pH relative to meat with a high pH. This may explain why the number of peaks per centimeter is decreased in meat with a low pH, but there might also be a direct effect of pH on fluorescence (which has not been investigated yet). If necessary, the effect of pH on fluorescence should be corrected by using other optical fibers in a probe to make ancillary measurements of light scattering.

Effect of Lipid

Lipid content in beef sometimes is weakly correlated with collagen content ($r \approx 0.44$) for reasons that are unclear at present. Fat is fluorescent because adipose cells are surrounded by Type III collagen fibers and are well supplied with blood vessels containing elastin. But the fluorescence is relatively weak and is similar to that of muscle fibers devoid of major connective tissues. From what is known of the optical properties of adipose tissue in relation to temperature (Chapter 10), one would expect the presence of adipose cells to enhance the penetration of UV and the transmittance of fluorescence, in a temperature-dependent manner (since transmittance of triglyceride increases with temperature). Thus, when the data are adjusted to a constant lipid content, the prediction of total collagen may be reduced. In practical terms, this means that other optical fibers in a probe should be used to measure lipid content using NIR.

CONCLUSION

As mentioned at the start of this chapter, the importance of connective tissue in the tenderness of top-quality beef is debatable. Arguing in support of the hypothesis, there is evidence that the abundance and heat stability of connective tissue influence beef toughness (Bailey, 1972; Dutson et al., 1976; Hall and Hunt, 1982; Light et al., 1985; Purslow, 1985; Burson and Hunt, 1986), and, as considered in Chapter 1, subjective assessments of animal age are used routinely in the beef-grading protocols of many countries. At present, animal age is assessed indirectly from the degree of carcass ossification (usually with little regard to mature frame size, breed differences, or rate of physiological maturation), primarily because older cattle produce tougher meat than younger cattle due to age-related increases in connective tissue resistance to cooking (Judge and Aberle, 1982). While generally true, there may be many exceptions, and biological variation may be responsible for some youthful animals producing tough meat (Wythes and Shorthose, 1991).

Anecdotal evidence carries far less weight than statistical proof but should not be ignored completely. Lawrie (1993) reported an exceptional case where a beef carcass with a collagen content six times higher than that of a half-sister carcass had exceptional toughness, while the sister carcass produced tender meat. Sporadic cases of extremely tough beef like this could be the cause of the steady erosion of customer confidence in beef quality (McDonell, 1988). Outliers in the general population, such as this case reported by Lawrie (1993), might be the tip of the iceberg that threatens the whole ship.

To all those critics of new methods for on-line meat evaluation who think that subjective evaluation of animal age and marbling is the best way to undertake a quality control program for beef tenderness, there is only one thing that can be said—keep chewing.

REFERENCES

Andersen, J. R., C. Borggaard, T. Nielsen, and P. A. Barton-Gade. 1993. *39th International Congress of Meat Science and Technology,* Calgary, Alberta, pp. 153–164.

Bailey, A. J. 1972. *J. Sci. Food Agric.,* 23:995.

Bailey, A. J. and T. J. Light. 1989. *Connective Tissue in Meat and Meat Products.* London: Elsevier Applied Science.

Bailey, A. J. and T. J. Sims. 1977. *J. Sci. Food Agric.,* 28:565.

Bashford, C. L. 1987. An introduction to spectrophotometry and fluorescence spectrometry, in *Spectrophotometry and Spectrofluorimetry,* D. A. Harris and C. L. Bashford, eds., Oxford: IRL Press, p. 15.

Bellon, G., A. Randoux, and J-P. Borel. 1985. *Collagen Rel. Res.* 5:423.

Bendall, J. R. 1967. *J. Sci. Food Agric.,* 18:553.

Bouton, P. E., F. D. Carroll, A. L. Fisher, P. V. Harris, and W. R. Shorthose. 1973a. *J. Food Sci.,* 38:816.

Bouton, P. E., F. D. Carroll, P. V. Harris, and W. R. Shorthose. 1973. *J. Food Sci.,* 38:404.

Bouton, P. E., A. L. Ford, P. V. Harris, W. R. Shorthose, D. Ratcliffe, and J. H. L. Morgan. 1978. *Meat Sci.,* 2:301.

Brownlee, M., H. Vlassara, A. Kooney, P. Ulrich, and A. Cerami. 1986. *Science,* 232:1629.

Burson, D. E. and M. C. Hunt. 1986. *J. Food Sci.,* 51:51.

Carroll, R. J., F. P. Rorer, S. B. Jones, and J. R. Cavanaugh. 1978. *J. Food Sci.,* 43:1181.

Davey, C. L. and K. V. Gilbert. 1975. *J. Food Technol.,* 10:333.

Davey, C. L. and K. V. Gilbert. 1976. *J. Sci. Food Agric.,* 27:244.

Davey, C. L. and A. E. Graafhuis. 1976. *J. Sci. Food Agric.,* 27:301.

Dransfield, E. 1992. *Meat Focus Internat.,* 1:237.

Dransfield, E. and D. N. Rhodes. 1976. *J. Sci. Food Agric.,* 27:483.

Dutson, T. R., R. L. Hostetler, and Z. L. Carpenter. 1976. *J. Food Sci.,* 41:863.

Etherington, D. J. 1984. *J. Anim. Sci.,* 59:1644.

Etherington, D. J. and T. J. Sims. 1981. *J. Sci. Food Agric.,* 32:539.

Fincato, G., F. Bartucci, M. Rigoldi, G. Abbiati, M. Colombo, O. Bartolini, M. L. Brandi, and V. De Leonardis. 1993. *J. Interdiscip. Cycle Res.,* 24:72.

Goll, D. E., W. G. Hoekstra, and R. W. Bray. 1964. *J. Food Sci.,* 29:608.

Goll, D. E., V. F. Thompson, R. G. Taylor, and J. A. Christiansen. 1992. *Biochimie,* 74:225.

Hall, J. B. and H. C. Hunt. 1982. *J. Anim. Sci.,* 55:321.

Harris, D. A. and C. L. Bashford. 1987. *Spectrophotometry & Spectrofluorimetry. A Practical Approach.* Oxford: IRL Press, p. 20.

Herring, H. K., R. G. Cassens, G. G. Suess, V. H. Brungardt, and E. J. Briskey. 1967. *J. Food Sci.,* 32:317.

Jeremiah, L. E., A. K. W. Tong, and L. L. Gibson. 1991. *Meat Sci.,* 30:97.

Judge, M. D. and E. D. Aberle. 1982. *J. Anim. Sci.,* 54:68.

Koohmaraie, M. 1992. *Biochimie,* 74:239.

Koohmaraie, M., G. Whipple, D. H. Kretchmar, J. D. Crouse, and H. J. Mersmann. 1991. *J. Anim. Sci.*, 69:617.

Kuhn, R. H., S. W. Peretti, and D. F. Ollis. 1991. *Biotechnol. Bioeng.*, 38:340.

Lawrie, R. A. 1993. *American Meat Science Association, Annual Reciprocal Meat Conference*, 46:1 – 8.

Li, J-K. and A. E. Humphrey. 1991. *Biotechnol. Bioeng.*, 37:1043.

Light, N., A. E. Champion, C. Voyle, and A. J. Bailey. 1985. *Meat Sci.*, 13:137.

Locker, R. H. and G. J. Daines. 1975a. *J. Sci. Food Agric.*, 26:1721.

Locker, R. H. and G. J. Daines. 1975b. *J. Sci. Food Agric.*, 26:1711.

Locker, R. H. and G. J. Daines. 1976. *J. Sci. Food Agric.*, 27:186.

Marsh, B. B. and W. A. Carse. 1974. *J. Food Technol.*, 9:129.

Martin, A. H., L. E. Jeremiah, H. T. Fredeen, and P. J. L'Hirondelle. 1977. *Can. J. Anim. Sci.*, 57:705.

McDonell, C. 1988. *Breeder and Feeder*, Ontario Cattlemen's Assoc., 191:14.

Miller, E. J. 1985. Recent information on the chemistry of the collagens, in *The Chemistry and Biology of Mineralized Tissues*, W. T. Butler, ed., Birmingham, UK: EBSCO Media.

Monnier, V. M. 1989. *Prog. Clin. Biol. Res.*, 304:1.

Monnier, V. M., R. R. Kohn, and A. Cerami. 1984. *Proc. Natl. Acad. Sci. USA*, 81:583.

Na, G. C. 1988. *Collagen Rel. Res.*, 8:315.

Nakano, T., J. R. Thompson, and F. X. Aherne. 1985. *Can. Inst. Food Sci. Technol. J.*, 18:100.

Odetti, P. R., A. Borgoglio, and R. Rolandi. 1992. *Metabolism*, 41:655.

Oimomi, M., Y. Kitamura, S. Nishimoto, S. Matsumoto, H. Hatanaka and S. Baba. 1986. *J. Gerontol.*, 41:695.

Ouali, A. 1992. *Biochimie*, 74:251.

O'Neill, I. K., M. L. Trimble, and J. C. Casey. 1979. *Meat Sci.*, 3:223.

Pearse, A. G. E. 1972. *Histochemistry. Theoretical and Applied, Vol. 2.* 3rd edit., Edinburgh: Churchill Livingstone.

Piller, H. 1977. *Microscope Photometry.* Berlin: Springer-Verlag.

Puchtler, H., F. S. Waldrop, and L. S. Valentine. 1973. *Histochemie*, 35:17.

Purslow, P. P. 1985. *Meat Sci.*, 12:39.

Rowe, R. W. D. 1974. *J. Food Technol.*, 9:501.

Schmitt, O., T. Degas, P. Perot, M-R. Langlois, and B. L. Dumont. 1979. *Ann. Biol. Anim. Bioch. Biophys.*, 19:1.

Schnider, S. and R. R. Kohn. 1981. *J. Clin. Invest.*, 67:1630.

Siano, S. A. and R. Mutharasan, R. 1991. *Biotechnol. Bioeng.*, 37:141.

Smith, S. H. and M. D. Judge. 1991. *J. Anim. Sci.*, 69:1989.

Suyama, K. and F. Nakamura. 1992. *Biorganic Med. Chem. Letters*, 2:1767.

Swatland, H. J. 1987a. *Histochem. J.*, 19:276.

Swatland, H. J. 1987b. *J. Anim. Sci.*, 64:1038.

Swatland, H. J. 1987c. *J. Food Sci.*, 52:865.

Swatland, H. J. 1987d. *J. Anim. Sci.*, 65:158.

Swatland, H. J. 1987e. *Meat Sci.*, 19:277.

Swatland, H. J. 1990. *J. Comput. Assist. Microsc.*, 2:125.

Swatland, H. J. 1991a. *J. Anim. Sci.*, 69:1983.

Swatland, H. J. 1991b. *Comput. Electron. Agric.*, 6:225.

Swatland, H. J. 1992. *Comput. Electron. Agric.*, 7:285.

Swatland, H. J. 1993a. *Food Res. Internat.*, 26:271.

Swatland, H. J. 1993b. *Food Res. Internat.*, 26:371.

Swatland, H. J. 1993c. *J. Anim. Sci.*, 71:2666.

Swatland, H. J., C. Warkup, and A. Cuthbertson. 1993. *Comput. Electron. Agric.*, 9:255.

Swatland, H. J., E. Gullett, T. Hore, and S. Buttenham. 1994. *Food Res. Internat.*, 28:23.

Swatland, H. J., T. Nielsen, and J. R. Andersen. 1995. *Food Res. Internat.*, in press.

Takahashi, K. 1992. *Biochimie*, 74:247.

Vishwanath, V., K. E. Frank, C. A. Elmets, P. J. Dauchot, and V. M. Monnier. 1986. *Diabetes*, 35:916.

Wang, N. S. and M. B. Simmons. 1991a. *Biotechnol. Bioeng.*, 38:907.

Wang, N. S. and M. B. Simmons. 1991b. *Biotechnol. Bioeng.*, 38:1292.

Wheeler, T. L., J. W. Savell, H. R. Cross, D. K. Lunt, and G. C. Smith. 1990. *J. Anim. Sci.*, 68:4206.

Whipple, G. and M. Koohmaraie. 1992. *J. Anim. Sci.*, 70:3081.

Wythes, J. R. and W. R. Shorthose. 1991. *Austral. J. Exp. Agric.*, 31:145.

Young, O. A. and T. J. Braggins. 1993. *Meat Sci.*, 35:213.

Zeiss, 1980. *Determination of Relative Spectral Fluorescence Intensities.* Bulletin A41-823.2-e. Carl Zeiss, Oberkochen, Germany.

Zimmerman, S. D., R. J. McCormick, R. K. Vadlamudi, and D. P. Thomas. 1993. *J. Appl. Physiol.*, 75:1670.

CHAPTER 10

On-line Assessment of Fat

INTRODUCTION

CHAPTER 1 CONSIDERED the on-line measurement of back-fat in relation to meat yield or cutability, but fat also is important for a number of other reasons. Yellowness, softness, and the smell of fat (boar taint) may detract from the overall quality of meat, while, according to tradition, marbling may make a positive contribution. This chapter reviews an assortment of attributes relating to fat that may be measured on-line.

When fat is abundant in meat, either as marbling or intermuscular fat, it modifies the overall reflectance and appearance of the meat. There is a direct effect: as the proportion of fat is increased in relation to the muscle, reflectance increases at many wavelengths until reaching the spectrum of 100% adipose tissue (Franke and Solberg, 1971). But there are also indirect effects of fat levels on meat color; for example, fat may retard oxygen penetration and metmyoglobin formation (Cutaia and Ordal, 1964).

ADIPOSE TISSUE

Fat or adipose tissue is composed of globular cells pressed tightly together. Mature adipose cells or adipocytes easily reach a diameter of about 100 μm and are almost filled by a single large droplet of triglyceride. Thus, the nucleus and cytoplasm of an adipose cell are restricted to a thin layer under the plasma membrane, which accounts for the low water content of fat. Mature adipose cells with very little cytoplasm contain few organelles. The large triglyceride droplet that fills most of the cell is not directly bounded by a membrane but is restrained by a cytoskeletal meshwork of 10-nm filaments, most conspicuous in the adipose cells of poultry. Adipose cells themselves are

259

kept in place by a meshwork of fine reticular (Type III collagen) fibers outside the cell. Large adipose depots usually are subdivided into layers or lobules by partitions or septa of fibrous connective tissue. In the layered subcutaneous fat of a pork carcass, the septa may follow the body contours, creating a weak boundary layer echo in the ultrasonic estimation of fat depth. Adipose depots are well supplied by blood capillaries, normally emptied by proper exsanguination, but often retaining a trace of hemoglobin that may give beef fat an amber tinge.

FAT COLOR

The yellow coloration of fat by β-carotene (Palmer and Eckles, 1914) is important in Canada, because the yellowness of beef fat is a cause for downgrading (Chapter 1). Unlike lean meat, in which reflectance changes very little from 0 to 30°C, the reflectance of adipose tissue is far higher at low temperatures than it is at high temperatures (Figure 10.1). Fiber-optic reflectance spectra of subcutaneous fat usually contain some trace of a Soret absorbance band and an oxyhemoglobin pattern at 542 and 578 nm, which, presumably, may be attributed to incomplete exsanguination of the dense capillary bed of adipose tissue.

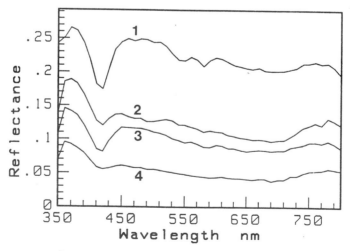

Figure 10.1 *Effect of temperature on the fiber-optic reflectance spectra of white fat at 0°C (1), pale yellow fat at 0°C (2), white fat at 40°C (3), and pale yellow fat at 40°C (4).*

The Soret absorbance band tends to be more conspicuous at a lower temperature (high reflectance) than at a high temperature (low reflectance).

Fat with a high β-carotene content has an overall lower reflectance intensity than white fat, particularly from 440 to 500 nm, where strong absorbance by β-carotene is separable from the Soret absorbance band of residual hemoglobin (Figure 10.2). In solution, β-carotene has a very low relative transmittance from 350 to nearly 500 nm (Swatland, 1987).

A point of technical interest shown in Figure 10.2 is that, in adipose tissue, a fiber-optic probe design with radiating fibers forming an annulus on a shaft (spectrum 4) gives the same shaped spectrum as a probe with parallel illuminating and receiving fibers (spectrum 3, also for white fat). Although the overall intensity of reflectance is different between spectra 3 and 4 (because of differences in standardization), the curvilinear shape is very similar and shows no evidence of a bias related to wavelength as may occur in muscle (Chapter 5).

In conclusion, the on-line measurement of fat color is relatively simple, with a choice of technology ranging from fiber optics to video. The critical point is that the temperature must either be kept constant or measured, because temperature has a strong effect on the overall reflectance of fat.

MARBLING

Video image analysis of marbling (Chapter 11) is likely to be the best method for marbling because, in a ribbed side of beef, the whole cross-sectional area of the rib-eye is exposed. This may give a much better sample of the marbling present in the whole muscle than is possible with a probe; however, there are many situations when it is commercially disadvantageous to rib the carcass, particularly for pork. Thus, it is of interest that fat-depth probes also detect marbling fat within the muscle, as indicated in the name of the Danish MQM probe — meat quality and marbling.

Another possibility for future on-line use is elastography (Ophir et al., 1994), using ultrasonic pulses to detect small movements caused by an external stress applied to the meat. This technique enables marbling and connective tissues to be visualized but not readily separated. At present, however, samples are removed from the carcass and measured in a large temperature-regulated water tank.

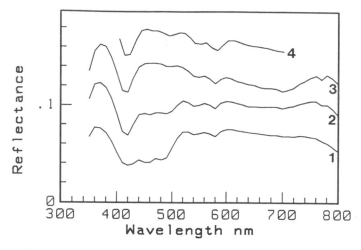

Figure 10.2 *Fiber-optic reflectance of subcutaneous beef fat. Spectrum 1 for lemon yellow fat, 2 for fat with a yellowish tinge, and 3 for white fat were measured with parallel illuminating and receiving optical fibers. Spectrum 4 for white fat was measured with radiating fibers forming an annulus on the shaft of the probe.*

Optical Probe

The relationship between optical window size, the size of the target tissue, and the shape of the resulting signal were considered in Chapter 3. Figure 10.3 shows some experimental results. It is difficult to build a series of probes with different sized optical windows, so this effect was simulated using a light guide with many optical fibers to form a window (Swatland, 1993). Then the size of the effective window was varied by altering the number of fibers used to make measurements. The effect of reducing the size of the optical window from approximately 2 to 0.5 mm diameter made the signal more irregular along the whole transect, so that the boundary between back-fat and the longissimus dorsi was less distinct, but a larger response was produced by the seams of marbling. Thus, the smaller optical window was more sensitive than the larger window to small anatomical irregularities, favoring the detection of marbling but not the detection of the fat to muscle boundary.

Steatosis

While on the subject of measuring marbling, a few words are needed on a condition called muscular steatosis, which often appears as an excessive degree of marbling (where the area of marbling greatly exceeds the area of muscle tissue). It is found most frequently in beef and pork but may also occur in mutton. There may be some indication of muscle abnormality before slaughter, such as an abnormal gait, but, usually, the condition is not found until the meat is cut. Excessive marbling fat and muscular steatosis sometimes overlap, and, often, it is only the restriction of muscular steatosis to a single muscle or muscle group in an otherwise poorly marbled carcass that makes it conspicuous. The causes of muscular steatosis are unknown, but strenuous muscle exertion may be involved, particularly in those muscles that are used when an animal rears up on its hindlegs as in mounting, so there might be a behavioral basis. The problem is notoriously sporadic. It may appear suddenly in carcasses from particular herds of animals, then disappear, thus making it very difficult to investigate scientifically.

SOFT FAT

Soft fat is a sporadic problem in many types of meat, but especially in pork, where it causes a variety of problems in meat cutting and marketing (Chikuni et al., 1982; Irie and Ohmoto, 1982; Dransfield and Jones, 1984). Separation of the back-fat from the underlying loin muscles

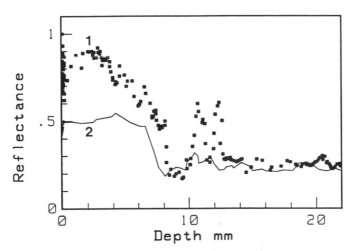

Figure 10.3 *Detection of fat-muscle boundary versus the detection of intramuscular marbling with small (1, signal shown by squares) and large (2, signal shown by line) optical windows.*

causes problems with automated equipment for removing the skin from pork loins, whereas, at the retail level, soft fat has increased translucency and a gray appearance that is unattractive to customers. In packs of sliced bacon, fat softness may cause the rashers to clump into a solid mass after slicing. Soft fat also is more likely to develop rancidity than hard fat.

The causes of soft fat are variable, ranging from nutritional increases in unsaturated fatty acids, notably linoleic acid, to inadequate carcass refrigeration. Thus, step one in an investigation is to check the deep temperature. If the cause has a biological basis in the fat, there are three on-line methods of measurement: ultrasonic, mechanical and optical.

A new aspect of soft fat in pork is the deliberate feeding of fish oil to increase the content of eicosapentaenoic and docosahexaenoic acids, thought to have a role in human health by reducing the risk of atherosclerosis and heart disease (Irie and Sakimoto, 1992). Effects on color are minimal, but softness and iodine number are increased.

On-line Measurement

On-line measurement is complicated by the fact that the back-fat on a pork carcass has three main layers (Moody and Zobrisky, 1966). The middle layer shows the greatest development when the back-fat is very thick and has the greatest proportion of extractable fat. The deepest layer contains the highest water content, and its thickness is correlated with marbling.

Ultrasonic measurement of soft fat is based on the fact that the velocity of ultrasound is greater in solid fat (V_S) than in liquid oil (V_L). Miles et al. (1985) proposed an acoustic parameter (ϕ) to determine the volume fraction of soft fat

$$\phi = \frac{1 - (V_L/V)^2}{1 - (V_L/V_S)^2}$$

The variation in ϕ explained 88% of the variance in penetrometer measurements of eighteen samples of back-fat ranging from hard to soft.

A handheld mechanical device (Figure 10.4) was developed by Dransfield at the Institute of Food Research at Bristol in the 1980s (Winstanley, 1987). A temperature sensor was used to adjust a mechanical measurement of softness, and it was used in a national survey of fat softness conducted by the UK Meat and Livestock Commission.

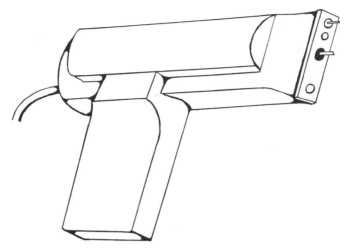

Figure 10.4 *Dransfield's handheld probe for the measurement of pork fat softness on-line (Winstanley, 1987).*

Optical properties of soft fat were evaluated by Irie and Swatland (1992). The softness of the inner layer of pork back-fat was assessed subjectively in the meat cooler at 4°C by palpation, using a four-point scale (1, soft; 2, slightly soft; 3, slightly hard; and 4, hard), like in the Japanese fat softness scores used for grading pork carcasses. Reflectance spectra obtained with a Colormet probe at 4°C for subjective fat softness scores 1 and 4 are shown in Figure 10.5. Both spectra have low reflectance around 420, 550, and 580 nm, probably from hemoglobin of residual erythrocytes in the dense capillary bed of adipose tissue. Reflectance is proportional to firmness, but optical separation is greatest between the softer levels of fat. In other words, slightly hard and hard fat had spectra that were not much different. However, the correlation of reflectance with softness is significant ($P < 0.01$) at all visible wavelengths over the full range from soft to hard. Soft fat has a higher refractive index and a lower melting point than hard fat.

The translucency of fat generally increases with temperature, with a corresponding decrease in reflectance; however, for pork back-fat, there is little change from 4 to 22°C, and not until 40°C does a decrease occur at all wavelengths. At 22°C relative to 4°C, there was a slight increase in the discrimination of subjective fat softness scores 3 and 4 (Figure 10.3). At 40°C, discrimination between subjective fat softness scores 3 and 4 was further enhanced, but with a loss of discrimination between subjective fat softness scores 2 and 3.

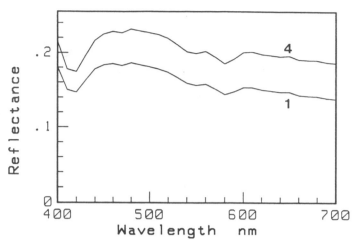

Figure 10.5 *Colormet fiber-optic reflectance spectra of soft (softness score 1) and hard (softness score 4) pork back-fat at 4° C.*

One might expect fat with a high proportion of unsaturated fatty acids to be more liquid, more translucent, and with a lower fiber-optic reflectance than fat with a high proportion of saturated fatty acids at the same temperature. Although basically correct, there appear to be some other factors involved as well, and the relationships of temperature, softness, and reflectance may be quite complex.

Carcass temperature decreases from around 40°C shortly after slaughter to around 4°C in a commercial meat cooler the day after slaughter. Both times are convenient for making probe measurements, either when the pork carcass is graded for fat depth shortly after slaughter or the next day, just before it is broken into wholesale cuts. At the earlier time, discrimination of subjective fat softness scores 3 versus 4 is difficult, while, at the later time, discrimination of scores 2 versus 3 is difficult optically. Unfortunately, it appears that the combined effects of degree of saturation and temperature may not be linear. Thus, there may be maximum and minimum temperature plateaus in reflectance, beyond or below which intrinsic differences in saturation have a negligible effect. At the reflectance minimum, when triglyceride is molten because of high temperature or unsaturation, there may be an independent background level of reflectance from cell membranes and other elements of adipose tissue microstructure. At the reflectance maximum, triglyceride may become opaque and almost white, thus concealing differences associated with fatty acid saturation. Also, the potential effects of adipose cell diameter are unknown.

In conclusion, optical methods might, perhaps, be suitable for grading the softness of pork fat, provided that the interactions of temperature and degree of saturation can be determined or cancelled. At present, the investment required to develop on-line methods for refractive index and melting point cannot be justified. Whether or not optical methods will prove superior to direct rheological measurement for on-line grading remains to be seen, but the advantage of being able to combine measurements of fat softness with other optical measurements of meat quality justifies further research on this topic.

BOAR TAINT

Young male pigs have a rapid rate of growth and high feed efficiency and produce lean carcasses, yet, in the United States and Canada, most of them are castrated, retarding their growth and encouraging fat deposition. Although there may be minor problems from soft fat, skin damage from fighting, and low curing yields with intact males, the primary reason for castration is the risk of boar taint in the fat.

Boar taint is an unpleasant odor that occasionally may become quite obnoxious when pork fat encounters a hot frying pan. Under typical commercial conditions in Europe, however, with pork from young males rather than old breeding boars, boar taint is only detectable by a small percentage of customers (Walstra, 1974). Thus, the problem owes its notoriety to pork obtained from old boars that have been used for breeding, and it need not be a problem in intact young males slaughtered at a relatively light weight. Keeping males intact for pork production provides a much larger base for genetic selection and avoids distress to the castrated animal.

The major cause of boar taint is the concentration of sex steroids in the fat (Patterson, 1968; Beery et al., 1969). Patterson (1968) identified the major factor as 5α-androst-16-ene-3-one, commonly called androstenone. Androstenone smells strongly of animal urine, but there are other testicular steroids in the 16-androstene family with a musk-like odor. Androstenone carried by the blood from the testes may accumulate in adipose tissue and parotid and submaxillary salivary glands (Gower, 1972). From the salivary glands, androstenone normally is transmitted as a pheromone in the boar's breath or saliva to the sow during mating

(Sink, 1967). Boar taint is a heritable trait (Jonsson and Wismer-Pedersen, 1974), although it may be suppressed immunologically (Claus, 1975) while retaining the greater leanness and growth efficiency of intact males (Brooks and Pearson, 1986; Brooks et al., 1986). Other causes of boar taint include skatole and indole, with a fecal odor, produced from the amino acid tryptophan in the gut (Hansson et al., 1980). Skatole accummulation may be controlled by a recessive gene (Lundstrom et al., 1994).

Off-line and On-line tests

Suspect carcasses may be tested with a hot iron, but the detection of volatiles by the operator is completely subjective (at present). Two objective methods have been developed, one for skatole and one for androstenone. Both methods require a sample to be removed from the carcass for analysis, but the methods are sufficiently rapid to be considered commercially.

A rapid, colorimetric test for skatole is used in Denmark (Mortensen and Sorensen, 1984), and a comparable test (colorimetric and suitable for automation) was developed in Canada by Squires (1990). The automated Danish skatole test may be run at 180 determinations per hour, using back-fat samples obtained forty to sixty minutes after slaughter (Andersen et al., 1993). Using an off-line test, data are available within twelve minutes, which is sufficient for sorting at the end of the chill tunnel. Data from 100,000 pigs are shown in Figure 10.6. The rejection point is at a level of 0.25 ppm skatole.

For androstenone, resorcylaldehyde and sulfuric acid in glacial acetic acid produce a purple color from androst-16-ene steroids and a pink color from cholesterol. Cholesterol must be removed with a digitonin affinity column from fat samples, but this is not required for samples of salivary gland because androstenone levels are higher. Results are correlated with sensory assessments by taste panels, $r = 0.85$ to 0.89 (Squires et al., 1991).

The skatole test is not highly regarded in some European countries, and there are plans to combine both methods to produce a test for both sources of boar taint (androstenone and skatole). However, Bonneau et al. (1992) concluded there was no advantage to testing for both androstenone and skatole.

On-line testing for boar taint may be possible in the future by using the Alabaster-UV semiconductor system (Berdagué and Talou, 1993).

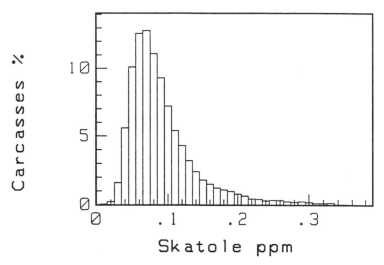

Figure 10.6 *Skatole determinations in 100,000 Danish pork carcasses from intact males (Andersen et al., 1993).*

The sensor is mounted in a stainless steel chamber with a UV source, and gases are exposed to a cycle of ozone and then flushed by air.

REFERENCES

Andersen, J. R., C. Borggaard, T. Nielsen, and P. A. Barton-Gade. 1993. *39th International Congress of Meat Science and Technology.* Calgary, Alberta, pp. 153–164.

Beery, K. E., J. D. Sink, S. Patton, and J. H. Zieglei. 1969. *J. Am. Oil Chem. Soc.,* 46:439A.

Berdagué, J. L. and T. Talou. 1993. *Sci. Aliments,* 13:141.

Bonneau, M., M. le Denmat, J. C. Vaudelet, J. R. Veloso Nunes, A. B. Mortensen, and H. P. Mortensen. 1992. *Livestock Prod. Sci.,* 32:63.

Brooks, R. I. and A. M. Pearson. 1986. *J. Anim. Sci.,* 62:632.

Brooks, R. I., A. M. Pearson, M. G. Hogberg, J. J. Perstka, and J. I. Gray. 1986. *J. Anim. Sci.,* 62:1279.

Chikuni, K., S. Ozawa, T. Koishikawa, and M. Yoshitake. 1982. *Bull. Natl. Inst. Anim. Ind., Chiba, Japan,* 38:115.

Claus, R. 1975. Neutralization of pheromones by antisera in pigs, in *Immunization with Hormones in Reproductive Research,* E. Nieschlag, ed., Amsterdam: North Holland Publishing Company, pp. 189–197.

Cutaia, A. J. and Z. J. Ordal. 1964. *Food Technol.,* 18:757.

Dransfield, E. and R. C. D. Jones. 1984. *J. Food Technol.,* 19:181.

Franke, W. C. and M. Solberg. 1971. *J. Food Sci.,* 36:515.

Gower, D. B. 1972. *J. Steroid Biochem*, 3:45.

Hansson, K-E., K. Lundstrom, S. Fjelkner-Modig, and J. Persson. 1980. *Swedish J. Agric. Res.*, 10:167.

Irie, M. and K. Ohmoto. 1982. *Jap. J. Swine Sci.*, 19:165.

Irie, M. and M. Sakimoto. 1992. *J. Anim. Sci.*, 70:470.

Irie, M. and H. J. Swatland. 1992. *Asian Austral. J. Anim. Sci.*, 5:753.

Jonsson, P. and J. Wismer-Pedersen. 1974. *Livestock Prod. Sci.*, 1:53.

Lundstrom, K., B. Malmfors, S. Stern, L. Rydhmer, L. Eliasson-Selling, A. B. Mortenssen, and H. P. Mortensen. 1994. *Livestock. Prod. Sci.*, 38:125.

Miles, C. A., G. A. J. Fursey, and R. C. D. Jones. 1985. *J. Sci. Food Agric.*, 36:215.

Moody, W. G. and S. E. Zobrisky. 1966. *J. Anim. Sci.*, 23:809.

Mortensen, A. B. and S. E. Sorensen. 1984. *Proceedings of the 30th European Meeting of Meat Research Workers* Bristol, p. 394.

Ophir, J., R. K. Miller, H. Ponnekanti, I. Cespedes, and A.D. Whittaker. 1994. *Meat Sci.*, 36:239.

Patterson, R. L. S. 1968. *J. Sci. Food Agric.*, 19:31.

Palmer, L. S. and C. H. Eckles. 1914. *J. Biol. Chem.*, 17:223.

Sink, J. D. 1967. *J. Theor. Biol.*, 17:174.

Squires, E. J. 1990. *Can. J. Anim. Sci.*, 70:1029.

Squires, E. J., E. A. Gullett, K. R. S. Fisher, and G. D. Partlow. 1991. *J. Anim. Sci.*, 69:1092.

Swatland, H. J. 1987. *Can. Inst. Food Sci. Technol. J.*, 20:383.

Swatland, H. J. 1993. *J. Anim. Sci.*, 71:2666.

Walstra, P. 1974. *Livestock Prod. Sci.*, 1:187.

Winstanley, M. 1987. *Institute of Meat Bulletin* (February): 10−13.

CHAPTER 11

Video Image Analysis

INTRODUCTION

THE INTRICACIES OF video image analysis (VIA) are the domain of the professional electrical engineer and computer scientist, but some knowledge of the principles is necessary to understand how VIA may be used on-line to measure marbling, muscle areas, and fat depth. Custom-built video systems are commercially available for the meat industry (Australian Meat Research Corporation, Sydney, Australia), but it is still an advantage to understand how they work. Back in 1978, when the USDA first commissioned VIA for carcass grading, the technology was an erudite speciality, and the first work was undertaken by NASA and the Jet Propulsion Laboratory (Cross et al., 1983). Now, the technology is easily available as an extra card for a personal computer, enabling us all to join in.

This chapter introduces the main components of hardware and software for the on-line evaluation of meat. Some of the other technologies described in this book may never get applied, while others that currently are in use may fade away. Like personal computers, however, VIA is here to stay and could emerge as the dominant noninvasive technology.

VIDEO HARDWARE

Pixels and Gray Levels

The basic components are a television camera providing a live video signal, usually at thirty frames per second, one frame of which is captured by a frame-grabber board in a personal computer. The frame-grabber board may record a square array of 512 × 512 pixels, but much larger arrays already are available, giving a higher resolution. A pixel

271

is one of the individual dots or controllable elements making up the whole picture. An array of 512×512 pixels is adequate at present for most operations in the meat industry because larger arrays require longer processing times, but, as personal computers get faster, larger arrays and higher resolution eventually will become cost-effective. In newer frame-grabber boards, the 4:3 width to height aspect ratio of the video image may be matched to an array, 640 pixels wide and 480 high.

With an eight-bit analogue to digital converter, each pixel may be set at one of 256 levels of light intensity (gray levels) from 0 to 255, although the lowest few usually contain electronic noise from the system itself and are not much use for analysis. Even an inexpensive frame-grabber board gives 256 gray levels, which is about eight or nine times more than we can separate with the human eye (although our eyes readily shift their whole dynamic range as we adapt from bright light to total darkness). A range of 256 gray levels is more than adequate for quantifying areas of muscle and fat but may not be adequate for absorbance measurements of meat pigments. The logarithmic nature of absorbance may place excessive reliance on just a few levels at the end of the gray scale, so more gray levels will give more accuracy. However, in most situations where the color of meat or fat is measured, this is not a problem, because we use only reflectance values, a straightforward ratio.

Storing video images may fill the memory of an ordinary personal computer very rapidly. Thus, the storage of images of rib-eyes or marbling on a videocassette recorder might seem like a good idea, enabling data collection from large numbers of carcasses in a meat cooler with portable equipment. But, usually, there is a considerable loss of detailed information from the picture, which may not be critical for rib-eye areas defined by the operator but may degrade performance for the automated measurement of marbling.

Vidicon Camera

There are two basic types of cameras available for video analysis of meat, the original vidicon tube and the newer CCD (charge-coupled device) cameras. In the vidicon tube, an image of the meat is focused on a small window at the end of the tube. The light of the image reacts with a photosensitive conductive coating on the inner side of the window, which is being scanned continuously in a raster (comb-like) pattern by an electron beam from the cathode at the other end of the tube. The image is transferred to the photoconductive coating as a pattern of high and low resistance regions and is converted to a stream of information in the current flow from the scanning beam. All this happens at very high speed, and the expertise that goes into designing a frame-grabber board essentially is how to latch this fast stream of information and store it, so that it can then be read by the personal computer, which is operating at a much slower speed than the television system. Never strip down a video camera in the laboratory to peek at the end of the vidicon tube, because dust is drawn in electrostatically: it may not show in normal operations but could spoil a future more demanding application, such as using the camera on a microscope.

One of the major disadvantages of an inexpensive or older vidicon camera may become apparent if it is used to record bright spots (such as a fireworks display at night). For each bright spot focused onto the outside of the vidicon screen, the corresponding highly activated spots on the inner photoconductive layer start to bloom or expand outwards by themselves. A large bright light may bloom to fill a large part of the screen. Thus, if spot sizes are being measured with VIA, they may appear larger than they should be, an unwelcome bias in measuring spots of intramuscular marbling fat.

It is a challenge for the television engineer to make the deflection and focusing coils of the vidicon camera create a perfect rectilinear scanning pattern on the photosensitive coating, so that the edges of the frame may be pushed inwards (pincushion effect) or bulge outwards (barrel effect). In attempting to define the $x: y$ coordinates on an image, like following the outline of a rib-eye, distortion from this artifact is unwelcome. In an inexpensive vidicon camera, the edge of the window also may be plagued by other problems, such as slight increases in thickness of the window glass that dim the image. In defining spots of marbling fat by their brightness, it may be wise to exclude those at the very edge of the screen.

Another problem with vidicon tubes may become apparent if the overall brightness of the image is changed, perhaps because of a change in ambient illumination or encountering a patch of DFD meat. This may cause the proportionality between the output signal and the brightness of spots of marbling fat to change across the whole picture. Again, if defining a fat to muscle boundary or the edges of marbling areas by their brightness, this is another unwelcome artifact. Avoid situations where a camera automatically changes its gain (as it adjusts automatically to different overall levels of brightness), because the new setting may have a different proportionality between brightness and the output voltage.

CCD Camera

Instead of a vacuum tube and electron gun, the CCD camera uses a chip with an array of photodetectors. The state of each detector in the array is changed by the photons it captures, until it is read and reset by scanning across the rows and down the columns, to produce an analogue voltage carrying the information of the whole picture.

CCD cameras have their own, but different, problems. Cramming thousands of photodetectors onto a square centimeter chip is a great feat of semiconductor engineering, but making each detector identical is a pursuit of perfection that is seldom achieved at a price we can afford in the meat industry. There is a tremendous difference in what we pay for apparatus in meat research, as compared to what astronomers pay, and most of us in meat research are working with a relatively low level of instrument performance. Differences in detector performance within the array on the chip show up as noise in the video signal. The sensitivities of the photosensitive coating in a vidicon tube can be matched to those in the human eye, but this is difficult with CCD cameras, which tend to be more sensitive to red and infrared light. If the visible image is focused properly, any infrared components are likely to be out of focus relative to the visible wavelengths, thus causing a fog.

Color Signals

Color is extremely important for operations such as automatically separating muscle from fat. With an inexpensive solid-state color camera, there may be a threefold loss in resolution if it is based on a monochrome chip using sets of three columns of photodetectors with filters for the three primary colors. Thus, we need a superior grade of CCD color camera for meat research.

The basic RGB (red, green, blue) color signal may be converted mathematically to other systems such as HSI (hue, saturation, and intensity), which relate more easily than RGB to our subjective impression of color (Russ, 1990). It is relatively simple to use HSI values to provide an interface to subjective terms of muscle and fat coloration but still difficult to obtain CIE color coordinates.

Standardization of RGB signals against black and white standards is extremely important. The emission spectra of illuminators change with the age of the source, as may the sensitivity of the camera. Thus, if the system is not properly standardized, the separation of muscle from fat on the basis of RBG levels may fail, giving erroneous results for muscle areas, marbling, or subcutaneous fat depth.

Illumination

There are a few general principles that facilitate the automated delineation of marbling fat and specific muscle areas such as the rib-eye. Lighting must be even and diffuse, which is usually achieved using reflector plates or fluorescent tubes (Cross et al., 1983) with a high degree of scattering around the illumination source. The angle of illumination should minimize the specular reflectance reaching the video camera. Specular reflection may be reduced still further by placing a sheet polarizer over the illuminator, with the transmittance axis in the same plane as the angle of incidence of the light on the meat. A further reduction of specular reflectance may then be achieved using another polarizer over the camera lens, with the transmittance axis perpendicular to that of the first polarizer.

A color signal is more useful than a monochrome signal for separating muscle from fat, and illumination for balanced color requires a high-intensity lamp likely to produce considerable heat. Thus, a heat-absorbing filter may be required immediately in front of the illuminator to protect both the sheet polarizer and the meat from heat. Removing infrared with the heat filter also may enhance the image if a CCD camera is used (the CCD camera may respond to infrared, which is likely to be out of focus relative to the visual image). Without these precautions, it is difficult to distinguish between specular reflectance and marbling.

IMAGE ANALYSIS SOFTWARE

Contrast Enhancement

One of the standard techniques of VIA is to pool all the pixels, regardless of their position in the image and plot them as a histogram, showing the number of pixels at each gray level. For example, in a monochrome image composed of closely related gray levels, the pixels will be grouped over a relatively small fraction of the range from 0 to 255, with none completely black (gray level 0) or white (gray level 255). By transforming the existing gray levels mathematically (subtracting the minimum gray level from the lower gray levels and adding to the higher gray levels), they can be spread over the whole dynamic range from 0 to 255. Then, when returned to their original positions in the image, the contrast will be greatly enhanced so that there are black shadows and white highlights.

With a color image represented by three separate 512 × 512 matrices for each of R, B, and G, another technique is to plot the pixels by their R versus G coordinates (regardless of their original position in the image). Video images of cut meat surfaces show two main clusters, a red cluster for areas of meat (myoglobin has absorbed most of the green light so that G is low) and a yellow-white cluster for fat (with higher G levels). Techniques for separating the two clusters depend on their degree of overlap. If they are completely separate, a diagonal across the graph separating the two clusters may be found very rapidly, but if the clusters overlap, a statistical cluster analysis technique may be required. Once this has been done, however, each pixel may be given a value of zero if it is from the cluster for red meat and a value of one if it is from the yellow-white cluster for fat and then returned to a 512 × 512 matrix using the original x:y coordinates. This creates a maximum contrast monochrome image where meat is black and fat is white. The binary numbers in such a matrix make efficient use of memory and may be processed rapidly using Boolean algebra. For example, an edge-finding algorithm may be used to move around the perimeter of a muscle area simply by moving ahead while keeping one edge of the leading point in contact with the meat.

Noise versus Marbling

Noisy images, like a television picture with white speckles, are all too common with inexpensive video equipment. Fortunately, there are some simple solutions, but a method of illustration is needed to help explain them. The figures that follow were obtained with a frame-grabber giving 256 gray levels for each pixel. Only part of the 512 × 512 matrix is used, so that the pixels are large enough to examine separately in the figures. To create black and white diagrams, the gray level of each pixel is converted into black and white stippling intensity, and the relative numbers of dots in the stippling pattern give the gray level. A form of contrast enhancement already has been done, spreading the gray levels of the meat over the whole dynamic range of the printer stippling patterns.

Figure 11.1 shows several square centimeters of heavily marbled beef rib-eye. The dark streaks are the marbling fat, because brightness gives a high gray level, and high gray levels give a high intensity of stippling.

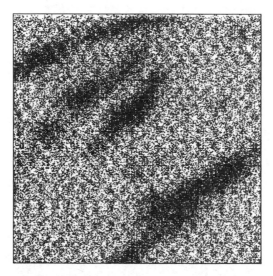

Figure 11.1 *Streaks of marbling fat from a rib-eye with moderately abundant marbling. The frame covers a few square centimeters in the center of the muscle. The brightness levels of the white marbling fat have been converted to video gray levels, which then have been converted to the probability of laserjet printer pixels being turned on.*

The distribution of marbling is seldom random: here, it is a series of parallel streaks. However, the use of black and white diagrams has inverted the appearance of the meat, but Figure 11.2 inverts the plotting logic, so that the white marbling fat now appears as clear white, while the red meat shows as black ink. Figure 11.2 also shows how the continuous image becomes segregated at the pixel boundaries. Figure 11.3 zooms into the top left corner of Figure 11.2 showing the pixel shape and gray levels more clearly.

Contour Mapping

Geographers mapping a landscape use topographical maps to show the heights and shapes of hills and mountains. Contour lines are drawn around features of the landscape at convenient separations of altitude. The same technique may be used to map the distribution of marbling, using gray levels in place of altitudes. Figure 11.4 shows a map of Figure 11.2 with contours at gray levels of 100 and 200: the marbling peaks are evident subjectively, but they are too irregular to be counted and measured automatically. Figure 11.5 shows the same marbling, but

Figure 11.2 *The same video image as in Figure 11.1 has been converted to a laserjet printer stipple pattern with seventeen degrees of filling, inverting the logic so that marbling appears white. Now the video pixels appear as rectangular blocks of uniform laserjet stippling.*

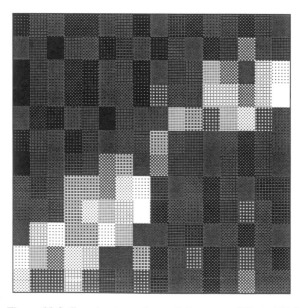

Figure 11.3 *Zooming in on the top left corner of Figure 11.3.*

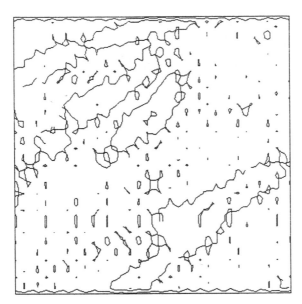

Figure 11.4 *A contour map of the video image in Figures 11.1 and 11.2, with the contour interval set at a gray level of 100. The major marbling sites are visible but not clearly enough to count or measure.*

Figure 11.5 *A repeat of Figure 11.4 with a higher contour interval set at a gray level of 150. The marbling peaks are visible, but their contours are broken.*

with a single contour at a gray level of 150: the map is simpler, but the contours of the marbling peaks are broken. The solution to this problem is to smooth the image to remove most of the noise.

Smoothing

Figure 11.6 takes the bottom row of pixels of Figure 11.2, which includes a streak of marbling at the midlength, and plots the gray levels at pixel positions across the scan. The spiky signal shows the high degree of noise between adjacent pixels (originating from the camera). Averaging successive pixels across the row gives a smooth line that better describes the light intensities on the original smooth meat surface. The streak of marbling that crosses the middle of the frame in Figures 11.1 and 11.2 now may be seen quite clearly in the smoothed signal. This method of neighborhood averaging may be expanded to process image matrices using a block pattern or kernel.

Neighborhood Averaging Kernels

Instead of averaging each data point with those immediately before and after it across the row, as in the smoothing of Figure 11.6, a simple kernel for two-dimensional smoothing may be used. Each pixel is averaged with the four adjacent pixels with which it shares a side. Even

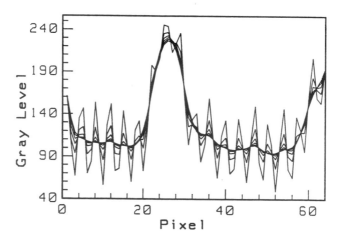

Figure 11.6 *The bottom row of pixels from the video image in Figure 11.2, plotting gray levels against x-axis pixel position. The spiky line is the noisy, raw signal, which is smoothed by successive passes of a moving average.*

this very simple method dramatically improves Figure 11.2 to give Figure 11.7, getting back the smoothness of the original in Figure 11.1, but without losing too much resolution. Figure 11.8 shows the advantage this produces for a contour map of Figure 11.7 (derived by smoothing Figure 11.2). Informative contour lines are neat and tightly spaced. Ignore the heavy black lines across the top and bottom rows: these are the contour lines at the start of the mapped area. Using the map in Figure 11.8, now it is possible to count and measure the marbling sites relatively easily. For example, the screen may be scanned in a raster pattern. When a marbling site is encountered, the area within the contour line may be found and then the whole area erased so that it cannot be counted again.

This simple method of neighborhood averaging may be summarized as a kernel

$$
\begin{array}{ccc}
0 & 1 & 0 \\
1 & 1 & 1 \\
0 & 1 & 0
\end{array}
$$

where the center of the kernel is the pixel being averaged with its four major neighbors: above, below, left, and right of the center. The corner

Figure 11.7 *A dramatic improvement of Figure 11.2 by the most simple, unweighted neighborhood averaging of pixels.*

Figure 11.8 *The greatly improved contour map of the processed image of Figure 11.7, allowing ten contour lines to be used, sufficient for a reasonably precise quantification of the marbling.*

positions are not used (0 weighting), and the others have an equal weighting (1). As the kernel is moved across the image so that each pixel in turn is at the center point of the kernel, the value at the center point is replaced by the average of five pixels in the pattern shown by the kernel. Note that, on the outermost rows and columns of the image, certain positions of the kernel are not available for averaging, so less pixels are averaged than for the internal rows and columns of pixels. When working row by row down an image, it is important to use the pixel intensities from the original or preceding image, and pixels that already have been averaged should not be reused. Thus, to reuse the existing matrix to store the newly processed image, a few extra rows (the height of the kernel) must be held in a memory buffer to avoid averaging data that have already been averaged.

From this simple start, many different types of more complex kernels are possible. The center pixel may be given more weight, thus reducing the chance of obliterating a one-pixel speck of marbling, as in the following example:

1 2 1
2 4 2
1 2 1

where the center pixel has a weighting of 4, the major neighbors 2, and the farthest neighbors 1. The weighted kernel is useful for preserving small specks of marbling or intermuscular fat that might get lost with smoothing. The contour map in Figure 11.9 shows a reasonable conclusion to the search for marbling in Figure 11.1, using the weighted kernel shown above, then turning pixels off or full on with the median gray level as a threshold. The relative numbers of off and on pixels give the overall area of marbling, while the numbers and sizes of the islands of marbling may be used to help predict the subjective response to the amount of marbling. Meat grading terminology (Chapter 1) uses subjective terms to describe marbling, and subjective impressions of marbling may be influenced by its distribution. Thus, the relationship between the numbers and sizes of marbling sites determined objectively by VIA may be quite complex. Although it is relatively simple to use the video data to generate simulated subjective responses, it makes more sense to short-circuit the whole process of subjective terminology and to use the video data to predict taste-panel responses to the cooked product.

The weighting of the center pixel may be increased for greater

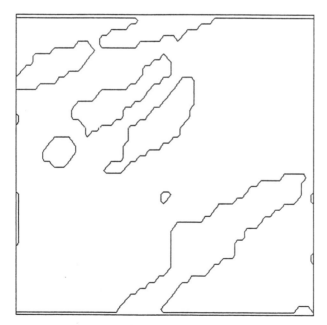

Figure 11.9 *A final decision on where the marbling is located in Figure 11.1, but another operator might have put the acceptance contour at slightly different positions, and both would have been misled by specular reflectance and irregular lighting at the meat surface.*

preservation of a single speck of marbling, but then it might be better to increase the size of the whole kernel to maintain the averaging effect of cancelling noise between pixels, such as

```
1   2   3   2  1
2   7  11   7  2
3  11  17  11  3
2   7  11   7  2
1   2   3   2  1
```

Complex Kernels

Much larger kernels than those shown above are used in commercial software, and the weighting of the center may follow a Gaussian pattern radially around the center. Another approach is to use a median filter in which, instead of replacing the center pixel by the arithmetic mean, the pixels of the kernel are ranked in order; then the median value is used to replace the center pixel. Sometimes median filters may be better than averaging filters because they maintain the brightness differences between pixels and cause less movement of boundaries, but simple averaging works well enough to detect marbling.

The kernels shown above may be adapted for sharpening boundaries or increasing the contrast between structures in an image, as in this Laplacian kernel.

```
-1  -1  -1
-1  +8  -1
-1  -1  -1
```

Not much happens as the kernel is moved across relatively smooth areas, but it produces a strong effect at boundaries and gives sharper images for the human eye. However, this is only because the eye is very responsive to boundaries, sketching in a cartoon of the image that we use for recognition, and the information content for our flat-visioned video analysis system may be no different or even lessened. Averaging kernels act as low-pass filters, letting through broad shapes but filtering out noise, whereas Laplacian kernels act as high-pass filters, ignoring broad shapes and emphasizing regional differences, which are easily influenced by noise. Thus, Laplacian kernels may be useful for sharpening the edges of a whole muscle but not for marbling.

VIA soon gets beyond the simple introduction attempted here, and readers are referred to Russ (1990) for an introduction to the subject. Once the kernels get large and sophisticated, the processing time for images increases accordingly.

APPLICATIONS

Kansas State University VIA

This pioneer system was relatively simple and measured total fat area, total lean area, number and area of marbling sites, and subcutaneous fat depth between ribs 12 and 13 (Cross et al., 1983). In tests to predict the composition of ribs 9 to 11 sections, inclusion of VIA lean area, fat area, and fat depth increased the coefficient of determination from 84% to 94%. Wassenberg et al. (1986) compared VIA to a three-person team of graders for predicting primal lean cutouts from sides of beef and obtained a slight improvement (coefficient of determination increased from 94 to 96%).

GLAFASCAN

The GLAFASCAN system for quantification of percent fat in cut meat moving on a black conveyer belt was announced in 1981 after four years of development by Unilever, Batchelor Foods, and the manufacturer, MicroMeasurements Ltd of Saffron Walden, Essex, UK (Winstanley, 1981; Anon., 1981). The GLAFASCAN is used to measure incoming frozen meat for canning and dehydration, scanning at 50 Hz and handling 3000 kg per hour with a 2-mm resolution of fat particles. Testing of the system is reported by Newman (1984a, 1984b).

Marbling

Marbling is particularly important in Japanese beef grading, where highly marbled meat sells at a considerable premium. Kuchida et al. (1992) used VIA to compute an index of marbling coarseness as the mean number of marbling pixels per marbling site. A form factor also was developed using the circumference of each marbling site. Breed differences and inheritable traits were demonstrated for Japanese Shorthorns versus Japanese Black cattle.

Muscle Areas

The area of the rib-eye or longissimus dorsi is used extensively in beef grading systems around the world as a primary determinant of lean meat yield. It is far from being a completely reliable predictor, but it is the best we have at present. From the first applications of VIA to carcass evaluation, it has been apparent that multiple sections may be required to develop accurate prediction equations (Butterfield et al., 1977). Beef animals, fortunately, have a constant number of vertebrae (relative to the variability of pigs and sheep), but variation in the length of vertebrae may occur, acting to weaken the value of the rib-eye area as a predictor of muscle mass. Thus, all other things being equal, cattle with long backs might have a small rib-eye area. The growth of the longissimus dorsi also is quite complex, and the height and width of the rib-eye may be partly independent. The longissimus dorsi has an irregular profile from a side view, as seen in a longitudinal dissection of the carcass, so that cross-sectional areas may vary along the length of the muscle. Corrections for these factors possibly could improve the predictive value of rib-eye areas.

With operator-defined muscle areas, VIA may be used to accelerate the acquisition of planimetry data on muscle and fat areas. Anada and Sasaki (1992) examined the area, circular length, long and short axis length, and center of gravity of several muscles at ribs 5 to 6. The most important variable for prediction of percent lean and percent fat was percent subcutaneous fat area; the best predictor of weight of lean and fat was the area of each tissue, and the distance between centers of gravity for the muscles was useful in predicting bone weight. Karnuah et al. (1994) found that eccentricity and orientation of the long axis of major muscles may also contribute to prediction equations.

Automated Rib-eye Measurement

The major technical problem in measuring rib-eye areas automatically is caused by the difficulty of separating adjacent muscles that touch the rib-eye, sometimes without any intermuscular fat separation. Several muscles are highly likely to have a muscle to muscle boundary with the longissimus dorsi, such as multifidus dorsi, longissimus costarum, and spinalis dorsi. When two intrinsically separate areas touch each other to produce an apparently continuous structure, sometimes they may be resolved separately by using two operations, erosion and dilation.

In the erosion operation, pixels are turned off all around the borders of the major areas, thus reducing the area. After several erosions, a crack separating two adjacent areas that touch may be completely opened. Once this is detected, the area is dilated by turning on the pixels around its edges, except where they would cause it to rejoin the area from which it has just been separated. Dilation is repeated until the areas are back to their original size, except in the seam where adjacent areas would touch. Thus, erosion and dilation may be used to create an imaginary seam of intermuscular fat between two adjacent muscles.

Small specks of intermuscular fat lined up along a true intermuscular separation facilitate opening a seam between the two adjacent areas in the video image. Thus, not only must this information be preserved if the image is smoothed to remove noise, but color information such as a high G value for a pixel may be used to enhance specks of fat, both for marbling determinations and opening seams. In some cases, as in the contiguous areas formed by muscles such as semimembranosus, adductor, and pectineus, separation may be completely impossible unless carcasses have a trace of intermuscular fat between these muscles.

In opening seams between contiguous muscle areas, it may be helpful to know the likely radius of curvature of the separation, by reference to the outline of the main muscle area that has already been established or by referring to a library of previous operator-guided separations. These two reference sources (the outline established so far and a library of reliable separations) also are useful for identifying separations that should not be pursued. The proper name for what we usually call the longissimus dorsi is longissimi thoracis et lumborum, indicating that this is a compound muscle composed of many small subunits. Sometimes a major crack through the muscle area is caused by the partial separation of one of the subunits, and knowing the likely angle and position of subunit separation may be used to prevent the erroneous subdivision of the whole compound muscle.

Erosion and dilation operations are available in most general purpose software packages for VIA. In a typical application, the operator writes a macro that goes through a series of erosions and dilations for the most difficult separation in the set of images to be processed; then the remainder of the images are processed automatically. But there is still considerable subjectivity in the operation: a macro that works satisfactorily for a series of fat carcasses with easy separations between muscles because of abundant intermuscular fat will fail if applied to images from leaner carcasses. For rib-eye areas, interactive decision making by the operator is almost essential for erosion and dilation methods, and complete automation of the rib-eye area measurement solely by erosion and dilation is unreliable.

It is difficult to penetrate the algorithms of commercial software after software compilation, but it is doubtful if any software has been written yet that uses artificial intelligence to recognize specific muscles of the carcass. This is bound to happen in the fullness of time, most likely as a by-product of medical imaging. But, at present, systems such as those available commercially from the Australian Meat Research Corporation still rely on the knowledge and hand-eye coordination of a human operator for a critical operation—locating the video camera at a specific point and orientation relative to the rib-eye area. This is facilitated by mounting the video camera on a rigid frame that the operator places against the rib and split vertebrae of the ribbed forequarter. Thus, the software may be written with the knowledge that the rib-eye area is somewhere within the video image at a constant position relative to the skeletal reference points.

If a specified muscle area such as the rib-eye is contained totally within a known part of the video frame and if the light intensity and camera response are regulated so that a constant threshold level may be used to distinguish muscle from fat, then this allows several methods to be used to locate the approximate center point of the specified muscle. The first step may involve a Boolean rejection of pixels below the threshold, regarding all the survivors as equal. The largest areas of contiguous surviving pixels (such as longissimus dorsi and spinalis dorsi) then may be skeletonized, progressively eroding the outer pixels of the area to converge on something resembling the center of gravity of the area. Then the muscle area may be reconstructed by flooding, turning on the peripheral pixels around the skeletonized areas, but not if this would involve linking up with another rebuilt area (Figure 11.10). Thus, the longissimus dorsi and spinalis dorsi areas, which might originally have been touching, now will be separated.

Another method of finding the rib-eye area is to drop a plumb line down the y-axis of the image after setting an acceptance threshold. The plumb line is lowered until it contacts an x-axis vector of on-pixels greater than a minimum value known for the muscle. In other words, this requires some prior knowledge of the minimum x-axis width of the specified muscle. Once inside the specified area, an edge-finding algorithm may be used to delineate the $x{:}y$ coordinates of the muscle perimeter.

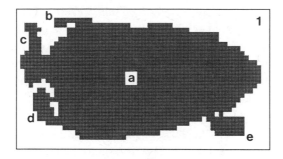

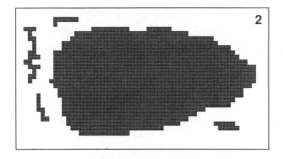

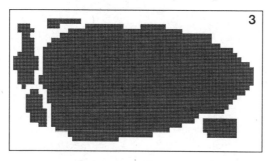

Figure 11.10 *(a) A rib-eye area formed by longissimus dorsi and surrounded by (b) trapezius, (c) spinalis dorsi, (d) multifidus dorsi, and (e) longissimus costarum. All the muscle has been set to one gray level, and pixel size has been exaggerated. Step 1 shows the original image, step 2 the eroded image, and step 3 the dilated image preserving the separations between contiguous muscles.*

Carcass Conformation

Provided that differences in bone length are taken into account, carcasses with bulging muscles have a higher relative meat yield than carcasses with sunken muscles. Thus, carcass conformation is important in subjective carcass evaluation. An increase in the girth of a deep muscle resulting from the radial growth of muscle fibers causes overlying superficial muscles to bulge outwards. Thus, the length of the superficial muscle is increased in a curvilinear manner as the deep muscle grows in size. When both fatness and carcass length are taken into account, a subjective appraisal of muscle conformation may be a useful guide to the anticipated lean yield of a carcass (Kempster and Harrington, 1980; Colomer-Rocher et al., 1980; Bass et al., 1981).

Using anatomical reference points on beef hindquarters, a triangle may be superimposed over the shape of the hindlimb, and the shape of the posterior profile of the hindlimb then may be quantified as an area relative to the area of the triangle (Bass et al., 1981). According to Kester (1991), VIA of carcass conformation is being used in beef grading in Denmark, but published information is hard to find. For VIA of muscularity, depth determination is enabled by illumination with a series of parallel planes of collimated light. On a flat surface this creates bright lines that are straight and parallel. But on the surface of a carcass, the lines curve to follow the contours of bulging muscle groups, enabling the software to estimate muscularity from bright-line deformation.

REFERENCES

Anada, K. and Y. Sasaki. 1992. *Anim. Sci. Technol.*, 63:846.

Anon. 1981. *Meat* (May):48.

Bass, J. J., D. L. Johnson, F. Colomer-Rocher, and G. Binks. 1981. *J. Agric. Sci., Camb.*, 97:37.

Butterfield, R. M., Y. Pinchbeck, J. Zamora, and I. Gardner. 1977. *Livestock Prod. Sci.*, 4:283.

Colomer-Rocher, F., J. J. Bass, and D. L. Johnson. 1980. *J. Agric. Sci., Camb.*, 94:697.

Cross, H. R., D. A. Gilliland, P. R. Durland, and S. Seideman. 1983. *J. Anim. Sci.*, 57:908.

Karnuah, A. B., K. Moriya, and Y. Sasaki. 1994. *Anim. Sci. Technol.*, 65:515.

Kempster, A. J. and G. Harrington. 1980. *Livestock Prod. Sci.*, 7:361.

Kester, W. 1991. *Beef* (Minneapolis), 28(4):14.

Kuchida, K., K. Yamaki, T. Yamagishi, and Y. Mizuma. 1992. *Anim. Sci. Technol.*, 63:121.

Newman, P. B. 1984a. *Meat Sci.*, 10:87.

Newman, P. B. 1984b. *Meat Sci.*, 10:161.

Russ, J. C. 1990. *Computer-Assisted Microscopy. The Measurement and Analysis of Images*. New York: Plenum Press.

Wassenberg, R. L., D. M. Allen, and K. E. Kemp. 1986. *J. Anim. Sci.*, 62:1609.

Winstanley, M. 1981. *Meat* (March):14−15.

CHAPTER 12

Cooking and Processing

INTRODUCTION

IN THE PRECEDING chapters, the main application for on-line meat evaluation has been for yield grading and detecting undesirable aspects of meat quality such as tough beef and PSE pork. On-line information may be used constructively to improve meat quality, as considered in Chapter 13, but it cannot be done rapidly. Meat processing, however, is more like a manufacturing industry, starting with raw materials and creating a new product, so that improvements in process control may offer an immediate commercial advantage. This chapter deals with the sensors that may be used to control and predict the quality of processed meat products.

Cooking is an important operation in meat processing, but it also has a much wider importance. Educating our customers so that they do not ruin the quality of the fresh meat we sell them is extremely important for the future of the whole meat industry. Restaurants and fast-food outlets already have a legal responsibility for meat cooking, and, with the general trend to make institutions responsible for things that once were the responsibility of the individual, the burden may one day be placed on the meat industry to make a demonstrable attempt to ensure that all meat is cooked appropriately by our customers. A preventive approach to problems tends to be more effective than a retroactive salvage operation, and we should be starting now in earnest to educate more of our customers, because procrastination creates a risk. In the domestic kitchen, many meat eaters already are overcooking their meat. If the trend continues, driven by a growing concern with microbial contamination of meat, it will further increase customer dissatisfaction with prime cuts of meat, because they are expensive yet toughened by overcooking. Other pressures are pushing our customers in this direction as well, for example, PSE pork. An exudate from PSE pork coagulating on the meat surface during cooking may encourage overcooking, until the meat surface looks dry, at which point the pork may be overcooked and tough.

Thus, at the research level, we need a major new initiative to improve our scientific understanding of the cooking process, not simplistic experiments that could have been done in the 1930s, but new technology dealing with chemical and physical changes produced by cooking. Experiments with on-line monitoring of cooking could help achieve this objective, enabling us to make cooking recommendations based on scientific fact, not opinion. For example, take the case of the internal color of a properly cooked fast-food hamburger. Thermal destruction of bacteria is related to pH and temperature, and temperature is related to the color change during cooking. Could on-line spectrophotometry become a useful tool to assess food safety?

ON-LINE MONITORING OF COOKING

Temperature

Technology for the on-line measurement of temperature is so well established and familiar that only a few words are needed here. A thermocouple is composed of two dissimilar electrical conductors joined at their ends to generate a thermoelectric voltage proportional to the temperature difference between the two end junctions. Iron-constantan and chromel-alumel thermocouples are the most common examples. Junctions may be exposed or covered and/or grounded. A thermistor is a resistive circuit component, usually a two-terminal semiconductor, whose resistance decreases as temperature increases. A Callendar's thermometer (platinum RTD probe) uses a thin film or wire of pure platinum, which has a relatively high resistance (for a metal), thus facilitating the measurement of the change in resistance with temperature.

Things to watch out for are the time constant, the length of time that the sensor takes to read the true temperature of the sample, and heat conduction along the sensor, its wire connections, and its protective shielding. For example, when monitoring the cooking of a roast, an unsuitable temperature probe may pick up heat from the air in the oven and conduct it into the roast faster than the heat can move through the meat. Thus, color changes in the interior of the roast may take place first along the route of the temperature probe. In making on-line measurements, one must also be careful about the length of wire between the sensor and the controller: when it exceeds the manufacturer's specifications, a signal transmitter may be required. Microwave cooking may be monitored using a temperature-sensitive fluorescent cap to a quartz optical fiber, using an optical controller rather like that shown in Figure 3.12. Infrared radiometers calibrated as thermometers are available for remote sensing of temperature.

Color Changes

It is generally agreed that meat with an appreciable myoglobin concentration changes from red to gray or grayish-brown when cooked (Pearson and Tauber, 1984). The brown pigments formed during cooking may include denatured globin nicotinamide hemichromes (Tappel, 1957), denatured myoglobin (Bernofsky et al., 1959), Maillard reaction products (Pearson et al., 1962), metmyochromogen (Tarladgis, 1962a), and hematin di-imadazole complexes (Ledward, 1974). The failure of meat to lose its red color when cooked is a persistent commercial problem, particularly with poultry, as with the nicotinamide-denatured globin hemochromes of cooked turkey rolls that appear pink (Cornforth et al., 1986).

Spectra 1 and 2 from Tappel (1957) in Figure 12.1 show, respectively, the macroscopic reflectance of raw and cooked beef measured with the reflectance attachment of a Beckman DR spectrophotometer. Spectra 3 and 4 from Swatland (1983) in Figure 12.1 show, respectively, the fiber-optic reflectance of cooked dark leg meat and white breast meat of chicken. Bearing in mind the differences in method of measurement, these data are reasonably consistent, and they show that cooking increases reflectance at most wavelengths, especially around 560 nm. Reduction, but not complete loss, of myoglobin absorbance around 560 nm also is apparent in spectra 5, 6, and 7 of Figure 12.1, which are data from Tarladgis (1962a) for cooked pork, lamb, and beef, respectively. These spectra were obtained with the reflectance attachments of a Beckman DK2 spectrophotometer but using an arbitrary calibration, which explains why reflectance is too high.

On-line Color Measurement

The color changes that occur during cooking are important for a variety of reasons. We rely exclusively on heat to protect us against

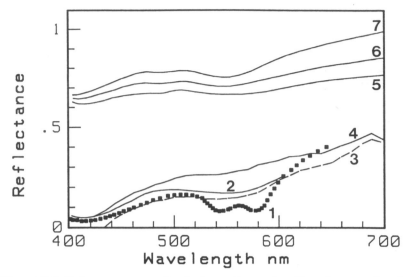

Figure 12.1 *Reflectance spectra of cooked meat. Spectra 1 and 2 are for raw and cooked beef, respectively (Tappel, 1957); spectra 3 and 4 are fiber-optic spectra for leg and breast meat, respectively, of cooked chicken (Swatland, 1983); and spectra 5, 6, and 7 for cooked beef, lamb, and pork, respectively (with arbitrary standardization from Tarladgis, 1962a).*

trichinosis and potentially deadly bacteria, but overcooking beyond the safe point may render meat unpalatable. In most practical situations, the safe point is judged not by temperature, but by color, which is why failure to attain a cooked meat color even after reaching a safe temperature is such a problem. On-line color measurement during cooking is relatively simple and may provide useful information to supplement that obtained by direct thermometry.

Figure 12.2 shows color changes during the cooking of lamb, in which the intermediate myoglobin concentration allows us to see cooking changes related both to light scattering and myoglobin denaturation. Samples were measured with a fiber-optic probe, using a central il-luminating fiber surrounded by six recording fibers leading to a grating monochromator (Swatland, 1989a). Cooking increases the reflectance at all wavelengths, but there are many subtleties that require more detailed analysis. Between 0 and 40°C, there are small changes in reflectance around the Soret absorbance band for myoglobin, but, for practical purposes, the color may be regarded as constant. Above normal body temperature, around 40°C, however, reflectance starts to increase, peaking at 70°C, and then decreasing as the temperature is increased

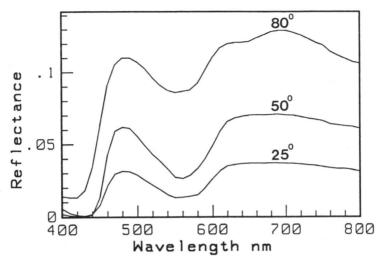

Figure 12.2 *Changes in fiber-optic reflectance during the cooking of lamb (Swatland, 1989a).*

still further. The data shown in Figure 12.3 are the means for ten samples.

For each wavelength, the slope for the change in reflectance per degree of temperature ($\Delta R/\Delta t$) is influenced by the magnitude of the initial reflectance, so that the slope is high for reflectance peaks and low

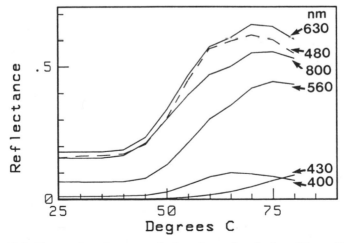

Figure 12.3 *Changes in reflectance of selected wavelengths from Figure 12.2 with respect to temperature.*

for valleys, which obscures what is really happening at different wavelengths. This unwanted effect is cancelled by dividing the temperature slope ($\Delta R/\Delta t$) for each wavelength by the mean reflectance for each wavelength at all temperatures, which we may call the adjusted temperature coefficient for each wavelength. Figure 12.4 shows the adjusted temperature coefficients, essentially the first derivatives of the data in Figure 12.3. The color changes related to myoglobin denaturation reach their maxima around 53 to 55°C, except for the Soret absorbance band wavelengths such as 430 nm. The Soret absorbance band has a spectral shift with temperature, and this obscures reflectance changes at nearby wavelengths. The practical significance of these spectral changes may be evaluated by using the weighted ordinate method to calculate CIE chromaticity coordinates, as shown in Figure 12.5. The loss of red myoglobin and the formation of new brown pigments that may occur within samples is masked by an effect of far greater magnitude, the increase in opacity that occurs as meat is cooked (Fox, 1987).

Gelatinization Birefringence

The effects of cooking on the microstructure of meat have been examined extensively by light and electron microscopy, and there is general agreement that the changes produced in collagen are particularly important (Birkner and Auerbach, 1960; Weidemann et al., 1967;

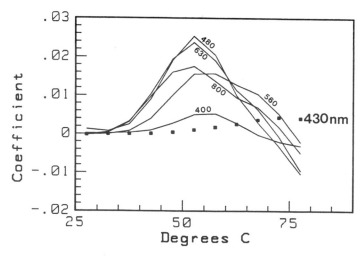

Figure 12.4 Adjusted temperature coefficients from Figure 12.3.

Schmidt and Parrish, 1971; Schaller and Powrie, 1972; Hegarty and Allen, 1975; Cheng and Parrish, 1976; Davey et al., 1976; Jones et al., 1977; Leander et al., 1980; Lewis, 1981; Purslow, 1985; Bailey, 1984; Bernal and Stanley, 1987). This is because bundles of muscle fibers are bound together by perimysium, and, when perimysial collagen is gelatinized by cooking, the bundles of muscle fibers are no longer restrained and much of the texture of the meat is reduced in strength. Although initially there were some uncertainties, it is now known that collagen fibers of the endomysium are also gelatinized during cooking (Swatland, 1975). Endomysial collagen fibers around individual muscle fibers are very important in understanding weight losses during cooking, because endomysial collagen fibers shrink as they are cooked, thus forcing out fluid from within the muscle fiber (Bendall and Restall, 1983).

The dominant component of perimysium is biochemical Type I collagen, but Type III collagen is also present as a minor component, and there are traces of Type V collagen as well (Light and Champion, 1984). In earlier research, the study of birefringence was important in helping to elucidate the molecular structure of collagen (Yannas, 1972; Chien, 1975; Wolman and Kasten, 1986), and polarimetry denaturation curves are still used in the study of collagen biochemistry (Condell et al., 1988). The first use of birefringence to study thermal changes in collagen was by Flory and Garrett (1958) who used polarized light microscopy,

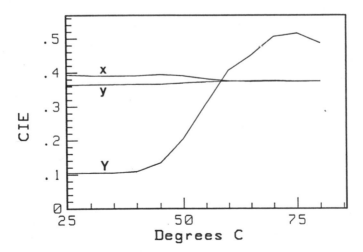

Figure 12.5 *CIE chromaticity coordinates (x and y) and luminosity (Y) estimated by the weighted ordinate method from spectra similar to those in Figure 12.2.*

although without quantitative measurements of path differences. Birefringence changes associated with the thermal denaturation of myofibrils were measured by Aronson (1966). The principles involved in the measurement of optical path differences in birefringent structures were introduced in Chapter 5, and the fundamental principles are explained by Hartshorne and Stuart (1970).

Birefringence changes during meat cooking may be measured in the laboratory using a hot stage on a polarized light microscope, so that thin sections of meat actually are cooked while they are being measured with the microscope. Obviously, this is not suitable as an on-line method for use in the meat industry, but some of the commercial equipment used in testing the polarization of optical fibers works on essentially the same ellipsometry principles. Thus, many of the measurements that are restricted to the laboratory at present, might be made on-line in the real world if the justification was sufficient for the investment in equipment. Another factor is that, within the last couple of years, there have been major advances in microscope ellipsometry. Instead of being restricted to a small part of the microscope field defined by an aperture and using analogue electronics, as in the results shown below (Swatland, 1990a), now it is possible to work on whole field images that have been digitized using a CCD camera.

Results obtained with retardation plates and a tilting compensator show that the slow optical axis of perimysial septa is parallel to their length when the septa are observed in transverse sections through the muscle. This information enables us to determine the correct orientation of the perimysium when the de Sénarmont method is used to measure path differences. The method of de Sénarmont (1840) for measuring path differences involves the conversion of elliptical polarized light to linear polarized light with a measurable azimuth, as developed by Goranson and Adams (1933). Although it is generally agreed that Type I collagen fibers are positively birefringent, the definition of the sign of birefringence appears to vary. Wolman and Kasten (1986) state that, "In an elongated crystal or structure the birefringence is positive in respect to length when the light velocity is higher in a plane parallel to the length of the structure." Bennett (1950), on the other hand, states that, "By convention, if the slow axis of transmission in a birefringent object corresponds to a given dimensional feature, the object is said to show positive birefringence with respect to that feature." For sections of perimysium taken transversely through a muscle, the latter definition is correct.

A further potential complication is that the perimysium is a complex structure composed of a meshwork of connective tissue fibers, associated cells, and intercellular matrix. In sections of perimysium taken transversely through a muscle, few of the fibers have their longitudinal axes exactly in the plane of sectioning, and the optical activity of Type I collagen fibers might conceivably be altered by the activity of Type III collagen fibers of the adjacent endomysium. It is important, therefore, that the angular arrangement of collagen fibers in meshworks does not alter the mean extinction position (Ortmann, 1975). In other words, bulk measurements of collagen birefringence obtained on-line will not be invalidated by this factor.

Line 1 in Figure 12.6 shows the birefringence changes during cooking of what we would expect to be a very strong, heat-resistant source of collagen, a perimysial collagen fiber from beef sternomandibularis muscle without any aging (two days postmortem). After 72°C, there is a gradual loss of birefringence as tropocollagen molecules lose their alignment and gelatinization occurs. Line 2 in Figure 12.6 shows the results from a collagen source that we would expect to be very weak and easily gelatinized.

Cooked turkey rolls sometimes fragment very easily when they are

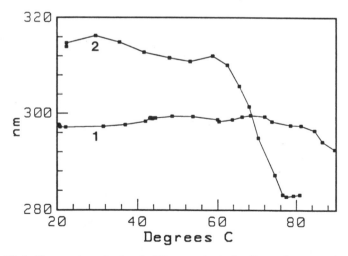

Figure 12.6 *Changes in optical path difference (nm) of collagen during cooking. Line 1 shows a perimysial collagen fiber from beef sternomandibularis muscle without aging (two days postmortem), and line 2 shows perimysium from a turkey breast muscle (several days postmortem) obtained from a commercial company that was having a problem with fragmentation of cooked turkey rolls.*

handled, so that they become difficult to display neatly on a delicatessen counter. Fragmentation occurs along the grain of the meat so that transverse slices fall apart, but, before processing, turkey breasts that are likely to do this may show a gaping fragmentation on the medial or dissected surface. On a radial basis, the growth of intramuscular connective tissues in turkeys does not keep pace with the growth of pectoralis muscle fibers during the earlier stages of growth (Swatland, 1990b), thus explaining why breast meat from heavier turkeys tends to be more tender than that from lighter birds (Grey, 1989). In other words, the meat gets more tender as the turkey gets older, the exact opposite of the condition we find in beef. Fragmentation of turkey rolls appears to have some relationship to the size of turkeys, and fragmentation seldom is apparent in breasts from small turkeys. Possibly, muscle fibers may outgrow their connective tissues, thus predisposing turkey breast muscles to fragmentation when they reach a large size. The final trigger to fragmentation may be the formation postmortem of large intercellular spaces filled by fluid released from the muscle fibers. Also, the collagen in turkey breast muscle fragmentation may be gelatinized very easily, as shown by line 2 in Figure 12.6. Starting at a relatively low temperature, 60°C, this particular sample of turkey meat showed a very rapid loss of molecular structure. The initial differences at 20°C between lines 1 and 2 in Figure 12.6 are unimportant and are caused by a thicker specimen for line 2. The significant features are the starting temperature for loss of birefringence and the rate of decline.

A major theoretical problem with this method concerns temperature-related changes in thickness of collagen fibers, because the destructive interference of ordinary and extraordinary rays is affected by the depth of the specimen, as well as the difference between its two refractive indices. In optical mineralogy, the thickness of geological specimens is seldom measured accurately in routine studies, and only an empirical value generally is used in calculations. Measuring the cooking of meat, the determination of the exact optical depth of perimysial boundaries is less of a problem because the main object of interest is the temperature-related change in path difference, not the path difference itself. However, results could be biased if collagen fibers were to show a substantial increase in diameter on thermal denaturation. But, fortunately, the bias is in the opposite direction to the experimental effect. In other words, if a collagen fiber contracts in length when heated so that its depth in the optical axis increases, one would expect the path difference to increase in trigonometric ratio. The fact that the path difference decreases, however, shows that the loss of birefringence is the dominant change.

The perimysium in cooked meat is important because it is where fractures start when meat is sheared or chewed (Purslow, 1985). In the many experiments that have been undertaken on meat cooking, a variety of results have been obtained on the temperature at which the perimysium undergoes thermal degradation. For example, Schmidt and Parrish (1971) only found degradation over 70°C; Cheng and Parrish (1976) found that degradation started at 70°C and was intense by 80°C; Leander et al. (1980) found that degradation started at 63°C and was complete by 73°C; and Bendall and Restall (1983) found that heat shrinkage started at 64°C and continued to 94°C. Although differences in experimental method may be involved, an important underlying biological basis may exist in the pattern of collagen cross-linking (Bailey, 1984). With differential scanning calorimetry, Judge and Aberle (1982) found that intramuscular collagen of older beef animals denatured at slightly higher temperatures than that of younger cattle and that the denaturation temperature decreased with postmortem conditioning. Biochemical differences in initial collagen structure and extent of postmortem degradation might be detectable as heat-related changes in the birefringence. It is not known at present whether a high proportion of heat-stable cross-links will cause a slow or a fast optical change after the start of thermal denaturation.

MEAT PROCESSING

Electrical Detection of Emulsifying Capacity

The emulsifying capacity of meat ingredients for further processing may be assessed by progressively adding oil during the formation of an experimental meat emulsion. The first oil to be added is trapped in a stable meat emulsion, but, finally, the emulsifying capacity of the system can stand no further additions and the emulsion breaks down. Webb et al. (1970) developed an electrical method to define the point of emulsion breakdown. Electrodes were located near the bottom of a vessel in which an emulsion was formed from salt-extracted meat protein solution by a mixing propeller. The placement of the electrodes produced a very sharp endpoint, with no change in the signal until the breakpoint was reached. The starting meat extract has a relatively low resistance that is not appreciably altered by adding droplets of oil, until the emulsion breaks and the oil insulates the electrodes.

A more dynamic view of these events may be obtained by using reverse logic, using impedance between the electrodes to assess the amount of oil that is being carried by the emulsion continually swirling between the electrodes (Figure 12.7). Bearing in mind that the purpose of an on-line sensor would be to avoid getting anywhere near an emulsion breakpoint, the impedance method might be more useful in practice, essentially measuring the dielectric properties of the meat mixture to assess the inclusion of fat.

Chopping

During the production of meat emulsions or batters, the reduction in particle size caused by chopping has little optical effect, but the incorporation of numerous bubbles has a major effect, making the product appear progressively more pale as it is chopped (Palombo et al., 1994). Thus, the lightness decreases if batters are stored and disproportionation occurs (gas diffusion from smaller to larger bubbles). Lightness was measured as L* with a Minolta CR-110, but an equivalent measure of scattering could be obtained in real time using an on-line fiber-optic system (Chapter 3).

Processed Meat Color

The color of processed meats is an important component of their overall commercial value. Thus, it is essential to keep the color of a

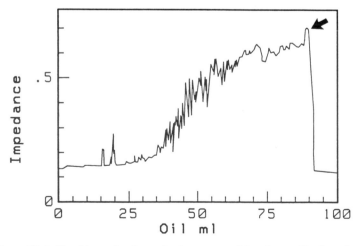

Figure 12.7 *Breaking point (arrow) of a meat emulsion detected by impedance.*

brandname product within specified color tolerances; otherwise, customers will object to what they consider an off color, even though it may be perfectly safe. Also, it is advantageous to produce step functions in the colors of different varieties of product, with premium products having the most attractive color. Color stability under typical retail display conditions is also important.

In meat curing, the action of nitrite and heat on myoglobin leads to the formation of dinitrosylhemochrome (Fox, 1987). Killday et al. (1988) have proposed that the eventual formation of nitrosyl-hemochromogen proceeds by reduction of several intermediates (nitrosylmetmyoglobin, to a radical cation, to nitrosylmyoglobin). Tarladgis (1962b) used a Beckman DK2 to measure the reflectance of cooked ham (spectrum 1 of Figure 12.8) and found that it had absorbance bands at 558 and 578 nm, resembling those of oxymyoglobin. No trace of this shape is evident by fiber-optic spectrophotometry with a 10-nm band width (spectrum 2 of Figure 12.8). Absorbance bands at 544 and 575 nm are quite distinct for nitric oxide myoglobin (Ginger and Schweigert, 1954). Possibly, these absorbance bands may be difficult to detect in intact tissue with a high overall intensity of reflectance due to scattering. Another possibility is that these absorbance bands may be lost after exposure to ambient illumination, as discovered in acetone extracts of nitroso-heme complex from cooked, cured pork (Hornsey, 1956).

Comparing the fiber-optic reflectance shown by spectra 2 and 3 in Figure 12.8, for cooked ham and raw pork, respectively, the major change associated with curing appears to be an increase in reflectance from 510 to 700 nm, with a loss of the myoglobin absorbance band around 560 nm. A similar change was detected for beef muscle processed to form salami when a coherent fiber-optic light guide was used to measure the interactance of the continuous phase of the salami between the fat particles (spectrum 4 in Figure 12.8).

Many processed meat products have a variegated appearance, a composite of chunks or specks of different ingredients. Thus, if individual parts of a variegated product are smaller than the aperture of a typical colorimeter, their color cannot be measured. This is not a problem with fiber optics, and separate measurements may be made for the continuous and discontinuous phases of the product (Swatland, 1985), as in the example shown for bologna (Figure 12.9). Tiny specks of bright color such as herbs and spices greatly enhance the appearance of an otherwise dull product and may be measured in situ with fiber optics (Swatland, 1985).

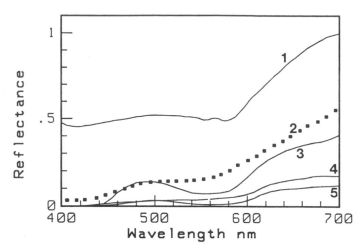

Figure 12.8 *Reflectance spectra of cured meats. Spectrum 1 is for ham measured with an arbitrary standardization (Tarladgis, 1962b); spectrum 2 shows the fiber-optic reflectance of ham standardized against barium sulfate (Swatland, 1985); spectrum 3 is the average for raw gluteus medius and iliopsoas pork muscles (Swatland, 1982); and spectra 4 and 5 were obtained with coherent fiber optics from the continuous phase of salami and raw beef, respectively (Swatland, 1985).*

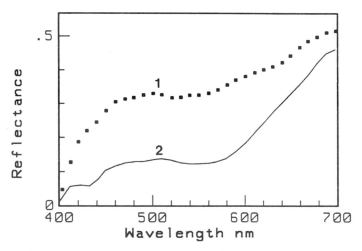

Figure 12.9 *Separation of colors in a variegated bologna, showing the light continuous phase (1) containing darker chunks (2).*

Prediction of Fat Content and Processing Loss

Meat quality sensors, robotics, and electronic control systems are becoming very important in the meat processing industry (Krol et al., 1988). Optical sensors may be used within either of two general strategies: either a peripheral sensor linked electronically to a central computer or a central sensor linked by fiber optics to a peripheral sample. The first option includes devices such as carcass fat-depth probes and portable fiber-optic probes for meat color that already are in use in the meat industry (Chapter 3). They tend to be dedicated, portable, relatively inexpensive, and ideal for use on batches of material, as shown in Plate 5. For the second strategy (a central sensor plus a long fiber-optic linkage), an extensive range of commercial equipment is available (Schirmer and Gargus, 1986). Infrared spectrophotometry is used for the analysis of many different food systems (Norris, 1984) and also may be used to monitor and control the functional properties of comminuted meat slurries used in meat processing. However, to understand the optical signals describing the functional properties of meat slurries, it is important to separate the effects of lipid content from those of pH-related protein functionality.

The data shown in Figures 12.10 and 12.11 are for a slurry of lean beef and pork fat (Swatland and Barbut, 1990). The lean and fat were comminuted separately for 1.5 minutes at 0 and 5°C in a bowl cutter,

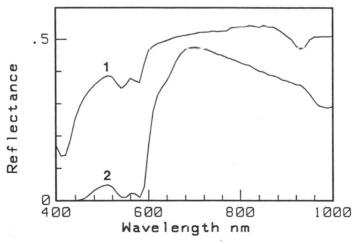

Figure 12.10 Fiber-optic reflectance spectra of starting ingredients, pork fat (1) and lean beef (2).

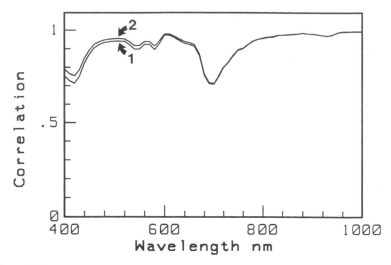

Figure 12.11 *Prediction of chemically determined lipid content (1) and cooking loss (2) in slurries of pork fat and beef lean.*

mixed in varying proportions, then reprocessed in the bowl cutter for thirty seconds. Sodium chloride was added to the meat slurries to extract salt-soluble functional proteins responsible for fluid retention and binding during heat processing. Salt concentration in the aqueous phase was kept constant at an ionic strength of 0.42, corresponding to a level of 2.5% NaCl and simulating typical commercial practice (Barbut and Findlay, 1989). The commercial properties of the meat slurries were evaluated using centrifugation fluid loss as a measure of water-holding capacity combined with fat mobility (Wardlow et al., 1973) and by finding the cooking losses (the amount of fluid fat, water, and solutes released during cooking in closed tubes).

The fiber-optic reflectance spectra of the starting ingredients are shown in Figure 12.10. Spectra for both lean and fat have evidence of Soret absorbance at 410–420 nm and a double indentation around 555 nm, possibly from oxyhemoglobin in the adipose tissue and oxymyoglobin in the lean meat, but they are widely separated. Thus, for a sensor located at the mixing point of two streams of ingredients (as in Figure 13.2), there are many wavelengths at which a simple feedback from an optical sensor could be used to control the mixture ratio.

For mixtures ranging from 100% lean to 100% fat, the corresponding chemically determined lipid contents ranged from 4.6% to 88.1%, respectively. Prediction of the lipid content is shown by line 1 in Figure 12.11, which closely follows line 2 for the prediction of cooking loss (which ranged from 6.9 for 100% lean to 74.4% for 100% fat), because cooking loss is strongly correlated with lipid content ($r = 0.99$, $P < 0.005$). Thus, in a continuous processing operation, it is possible to control both the fat content, relative to nutritional labeling, as well as the cooking losses.

Prediction of pH and Processing Loss

Processing losses are determined by pH, as well as by fat content, so it is important to understand how these two factors interact (there is no pH effect in the data shown in Figures 12.10 and 12.11 because all the lean beef was from a single source and thoroughly mixed). Effects related to pH show up most clearly for pork and poultry meat. With a typical range from mildly PSE to mildly DFD pork (but little range in fatness, from 2.0 to 6.0%), pH is correlated with both fluid loss from the raw meat and loss during cooking (lines 1 and 2 in Figure 12.12). However, if data are adjusted to a constant lipid content (3.42%), the correlations of internal reflectance with fluid loss and with cooking loss are reduced in strength (lines 3 and 4 in Figure 12.12).

Interaction of pH and Fat Content

For the effect of fat content (Figure 12.11), the pattern of cooking losses is caused by the reduction in protein content (from 21.7 to 0.0%) as the lipid level increases (from 4.6 to 88.1%). Reduced protein content causes an increase in cooking loss because protein is the major functional component responsible for fluid retention (Hermansson, 1985; Hamm, 1986; Whiting, 1988). The protein content of processed meat products ranges from about 22% for a lean muscle product, such as oven roasted turkey breast, down to the minimum legal level (11% in Canada). But a certain amount of high-quality protein is required to form an acceptable cohesive meat product.

Near-infrared reflectance is used in many commercial operations to predict the fat content of meat (Norris, 1984), but similar results may be obtained through optical fibers on-line. Thus, when near-infrared internal reflectance is high, a slurry of comminuted meat is likely to have a high fat content, leading to high fluid and cooking loss during further processing; however, there are factors other than lipid that affect the commercial properties of meat slurry, such as the pH-related functional state of myofibrillar proteins (Figure 12.12) and the content of connective tissue proteins. Near-infrared reflectance measured through optical fibers is negatively correlated with pH. Thus, the correlations of reflectance with lipid content and with pH are additive. This is a fortunate coincidence that strengthens the reliability of fiber-optic measurements for the prediction of fluid loss and cooking loss on-line. If relationships were subtractive rather than additive, the technology would be prone to failure.

Prediction of Connective Tissue and Processing Loss

Lipid content and pH-related aspects of protein functionality of a meat slurry may be measured via optical fibers, as described above, but, in many situations, the results are improved if some consideration is also given to connective tissue content. This is especially important when poultry skin is used as a low-cost filler in processed poultry products but may have other applications, as in automatically controlling the levels of modified connective tissues to improve the quality of low-fat frankfurters (Eilert et al., 1993). In excess, skin in poultry products causes a deterioration of product quality, so that the level must be controlled to maintain a balance between least-cost formulation and product quality. Carcass connective tissues may be measured by fluorometry, as described in Chapter 9. The same technology may be applied on-line in poultry processing, although the emphasis is changed

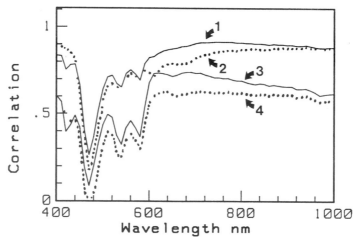

Figure 12.12 Prediction of fluid loss (1), cooking loss (2), fluid loss adjusted to constant lipid content (3), and cooking loss adjusted to constant liquid content (4) for a processing slurry of pork.

from measuring distribution patterns of small elements of connective tissue, to measuring overall, integrated fluorescence.

A quartz-glass rod, one or two centimeters in diameter and any reasonable length, may be coated with aluminum on its sides to give an internally reflecting surface. In continuous processing, one end of the rod may be mounted through a Delrin pipe fitting to be in contact with the meat slurry, while the other end connects to the optical apparatus (similar to that in Figure 3.12). The surfaces at both ends of the rod are polished, and the end in contact with the meat slurry must be self-cleaning: a static accumulation of product over a recessed optical window will totally defeat the purpose of the sensor. Working with a large-diameter quartz rod, instead of an optical fiber, makes it possible to use an ultraviolet pencil lamp (a miniature fluorescent tube) as the illuminator. Relative to arc lamps, pencil lamps are relatively inexpensive and have a long operational life. Light from fluorescent tissues in contact with the distal end of the quartz rod passes through the dichroic mirror, through a grating monochromator, through a stray light filter, through a solenoid shutter, and onto a photometer. The grating monochromator is required only at the research level: once suitable wavelengths for the product have been identified, the on-line sensor is far simpler and has only a fixed filter for the wavelengths required.

For the data shown in Figures 12.13 and 12.14, from Swatland and

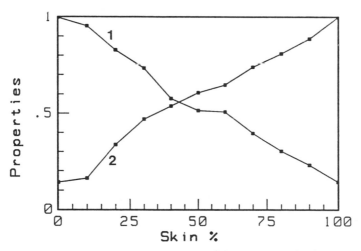

Figure 12.13 *Effect of skin content on the rheological gel strength (line 1, maximum 29 N penetration resistance) and cooking loss (line 2, maximum 12%) of a slurry composed of chicken breast and skin.*

Barbut (1991), chicken skin and muscle were comminuted separately for 1.5 minutes at 0 and 5°C in a bowl cutter and muscle-skin mixtures were recomminuted with the addition of salt in the bowl cutter for thirty seconds. Muscles contained 75.0% moisture, 22.1% protein, 1.6% lipid, and 1.3% ash, while skin contained 47.9% moisture, 45.1% lipid, 6.6% protein, and 0.4% ash. The collagen content of skin and muscle was determined from the hydroxyproline content, using overnight hydrolysis in 6N HCl, oxidation with chloramine-T, and spectrophotometry with p-dimethylaminobenzaldehyde. Hydroxyproline content may be converted to total collagen by multiplying by 7.25 (Goll et al., 1963). Collagen concentrations of the muscle and skin slurries were 5 and 30 mg/g, respectively, close to average values (Nakamura et al., 1975; Satterlee et al., 1971). The NaCl concentration in the muscle slurry was 1.75%, corresponding to a concentration of 2.25% in the aqueous phase.

The proportion of skin in a product is negatively correlated with rheological gel strength and positively correlated with cooking loss, as shown in Figure 12.13. Chicken meat slurry has only relatively weak fluorescence, whereas skin slurry has relatively strong fluorescence peaking at 470 nm. Muscle-skin mixtures have intermediate intensities of fluorescence. Fluorescence intensity is correlated positively with percent skin and cooking losses and negatively with gel strength, but the

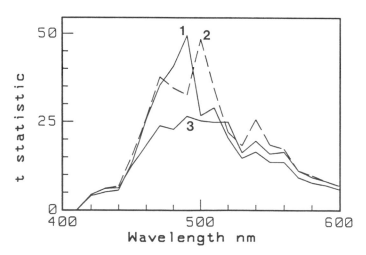

Figure 12.14 *Spectral distribution of the absolute value of the t-statistic for the correlation of fluorescence emission with skin percent (1), gel strength (2), and cooking losses (3) in slurries of chicken breast meat and skin.*

correlations are almost congruent and uniformly strong ($r > 0.9$) from about 450 to 560 nm. In a situation such as this (where all the correlations are strong across a wide region of the spectrum), the t-statistic of the regression is more useful than the coefficient of correlation for choosing a fixed wavelength at which to measure fluorescence emission on-line, as shown in Figure 12.14.

Fluorometry may be more useful than reflectance measurements for monitoring the skin content of poultry products. Acton and Dick (1978) examined cooked poultry loaf (dark meat) with a skin content from $0-50\%$. No effects on Gardner L (lightness) or b (yellowness) values were detected and only a slight decrease in the a value (redness) was detected with increased levels of skin. In conventional poultry products, it is unlikely that a manufacturer would add skin at levels much over 15 or 20%. Within the range from $0-20\%$, there is a strong effect of skin content on gel strength and total cooking loss. And within this range, differences in fluorescence are readily detectable and may be used for monitoring product composition or feedback control of product composition (Chapter 13).

In a practical application, there are bound to be problems that do not show up in the laboratory proving of the technology. In this case, precautions would be needed in the exclusion of ambient light and ensuring that the slurry is pressed uniformly against the quartz window. Also, it might be necessary to consider the effect of myoglobin (from red muscle) and hemoglobin (from hemopoietic tissue in any mechanically deboned meat added to the product). Ultraviolet light has poor penetration into meat slurry, because scattering is inversely proportional to a power of wavelength, as considered in Chapter 5, and the fluorescence measured is dominantly from the surface layers of the slurry. However, a strong chromophore such as myoglobin in these layers could absorb some of the fluorescence. Thus, in a product with a variable level of pigmentation, it might be necessary to monitor the pigmentation separately using another sensor and to use this information to adjust the fluorescence data. The collagen content of chicken muscle may also vary between muscles and with animal age (Nakamura et al., 1975).

Prediction of Structural Integrity

Structural integrity is a vital aspect of quality in all meat products intended for thin slicing, because a product that fragments too easily may become difficult to display, package, or utilize. In many situations where meat fragments are held together by a heat-set meat emulsion or by gelatin, integrity may be controlled using conventional meat processing knowledge, but there are some situations where integrity is difficult to control, such as in the fragmentation of cooked turkeys rolls.

The expansion of turkey production in recent years has been greatly facilitated by the increased amount of product that has been secondarily processed, rather than being consumed as a traditional roast. Turkey breast muscles show considerable variation in their processing characteristics (Grey, 1989) because of differences in connective tissue content (Swatland, 1990b) and pH (Vanderstoep and Richards, 1974; van Hoof, 1979; Barbut, 1993). Biological sources of variation include body size (Stiles and Brant, 1971; Swatland, 1989b; Maruyama et al., 1993), sex (Clayton et al., 1978), muscle pathology (Sosnicki and Wilson, 1991), genetics (Johnson and Asmundson, 1957; Hayse and Moreng, 1973), nutrition (Grey et al., 1986), environment (Halvorson et al., 1991), and behavior (Noll et al., 1991). These interact with postproduction differences in capture, transport, slaughter, and carcass cooling (Ma and Addis, 1973; Ngoka et al., 1982; Raj, 1994). Meat processing problems are caused by a high degree of variability or overall low quality of the raw meat, leading to poor slicing characteristics, toughness, or low moisture retention in products such as oven-roasted turkey breast. Slicing characteristics and toughness problems reach as far as the customer, whereas moisture binding problems require the processor to rebag the product if too much fluid is released in the cook-in-bag. All these conditions have an adverse affect on profitability.

The effect of variation in raw material on endproduct quality was determined in a batch of turkey breast muscles selected by a commercial processor to cover the maximum range of raw meat quality normally encountered (Swatland and Barbut, 1995). The processing characteristics of the samples were measured in the laboratory, simulating industrial conditions. The objective was to search for optical predictors of finished product quality. Using a UV connective tissue probe, the frequency of fluorescence peaks in raw samples was correlated with the maximum force required for penetration of the cooked product ($r = 0.74$, $P < 0.005$) and with Young's modulus ($r = -0.71$, $P < 0.005$), while the mean height of strong fluorescence peaks was correlated with the tensile strength of the cooked product ($r = 0.72$, $P < 0.005$). NIR birefringence was correlated with the water-holding capacity of raw samples ($r = 0.85$, $P < 0.0005$) and with fluid losses during cooking

$(r = -0.82, P < 0.005)$. The UV probe was a better predictor of final product structure than either pH or paleness measured with a colorimeter. NIR birefringence was almost as useful as pH and better than colorimeter paleness for predicting water-holding capacity and cooking losses.

REFERENCES

Acton, J. C. and R. L. Dick. 1978. *Poultry Sci.*, 57:1255.

Aronson, J. F. 1966. *J. Cell Biol.*, 30:453.

Bailey, A. J. 1984. The chemistry of intramolecular collagen, in *Recent Advances in the Chemistry of Meat.*, A. J. Bailey, ed., London: Royal Society of Chemistry.

Barbut, S. 1993. *Food Res. Inter.*, 26:39.

Barbut, S. and C. J. Findlay. 1989. *CRC Critical Reviews in Poultry Biology*, 2:59.

Bendall, J. R. and D. J. Restall. 1983. *Meat Sci.*, 8:93.

Bennett, H. S. 1950. The microscopical investigation of biological materials with polarized light, in *McClung's Handbook of Microscopical Technique*, R. McClung Jones, ed., New York: Hoeber, p. 635.

Bernal, V. M. and D. W. Stanley. 1987. *Can. Inst. Food Sci. Technol. J.*, 20:56.

Bernofsky, C., J. B. Fox, and B. S. Schweigert. 1959. *Food Res.*, 24:339.

Birkner, M. L. and E. Auerbach. 1960. *The Science of Meat and Meat Products*. San Francisco, CA: W. H. Freeman and Company, p. 24.

Cheng, C. S. and F. C. Parrish. 1976. *J. Food Sci.*, 41:1449.

Chien, J. C. 1975. *J. Macromol. Sci. Revs. Macromol. Chem.*, C12:1.

Clayton, G. A., C. Nixey, and G. Monaghan. 1978. *Br. Poult. Sci.*, 19:755.

Condell, R. A., N. Sakai, R. T. Mercado, and E. Larenas. 1988. *Collagen Rel. Res.*, 8:407.

Cornforth, D. P., F. Vahabzadeh, C. E. Carpenter, and D. T. Bartholomew. 1986. *J. Food Sci.*, 51:1132.

Davey, C. L., A. F. Niederer, and A. E. Graafhuis. 1976. *J. Sci. Food Agric.*, 27:251.

de Sénarmont, H. 1840. *Annls. Chim. Phys.*, 73:337.

Eilert, S. J., D. S. Blackmer, R. W. Mandigo, and C. R. Calkins. 1993. *J. Muscle Foods*, 4:269.

Flory, P. J. and R. R. Garrett. 1958. *J. Am. Chem. Soc.*, 80:4836.

Fox, J. B. 1987. The pigments of meat, in *The Science of Meat and Meat Products*, J. F. Price and B. S. Schweigert, eds., Westport, CT: Food & Nutrition Press, Inc., pp. 193–216.

Ginger, I. D. and B. S. Schweigert. 1954. *J. Agric. Food Chem.*, 2:1037.

Goll, D. E., R. Bray, and W. G. Hoekstra. 1963. *J. Food Sci.*, 28:503.

Goranson, R. W. and L. H. Adams. 1933. *J. Franklin. Inst.*, 216:475.

Grey, T. C. 1989. Turkey meat texture, in *Recent Advances in Turkey Science*, C. Nixey and T. C. Grey, eds., London: Butterworths.

Grey, T. C., N. M. Griffiths, J. M. Jones, and D. Robinson. 1986. *Lebensm. Wiss. Technol.*, 19:412.

Halvorson, J. C., P. E. Waibel, E. M. Oju, S. L. Noll, and M. E. el Halawani. 1991. *Poult. Sci.*, 70:935.

Hamm, R. 1986. Functional properties of the myofibrillar system and their measurement, in *Muscle as a Food*, P. J. Bechtel, ed., New York: Academic Press, p. 135.

Hayse, P. L. and R. E. Moreng. 1973. *Poult. Sci.*, 52:1552.

Hermansson, A. M. 1985. Water- and fat-holding, in *Functional Properties of Food Macromolecules*, J. R. Mitchell and D. A. Ledward, eds., New York: Elsevier Applied Science Publishers, p. 273.

Hartshorne, N. H. and A. Stuart. 1970. *Crystals and the Polarising Microscope.* London: Edward Arnold, p. 309.

Hegarty, P. V. J. and C. E. Allen. 1975. *J. Food Sci.*, 40:24.

Hornsey, H. C. 1956. *J. Sci. Food Agric.*, 7:534.

Johnson, A. S. and V. S. Asmundson. 1957. *Poult. Sci.*, 36:1052.

Jones, S. B., R. J. Carroll, and J. R. Cavanaugh. 1977. *J. Food Sci.*, 42:125.

Judge, M. D. and E. D. Aberle. 1982. *J. Anim. Sci.*, 54:68.

Killday, K. B., M. S. Tempesta, M. E. Bailey, and C. J. Metral. 1988. *J. Agric. Food Chem.*, 36:909.

Krol, B., P. S. van Roon, and J. H. Houben. 1988. *Trends in Modern Meat Technology 2.* Wageningen, the Netherlands: Pudoc.

Leander, R. C., H. B. Hedrick, M. F. Brown, and J. A. White. 1980. *J. Food Sci.*, 45:1.

Ledward, D. A. 1974. *J. Food Technol.*, 9:59.

Lewis, D. F. 1981. *Scanning Electron. Microscop.* (III):391.

Light, N. and A. E. Champion. 1984. *Biochem. J.*, 219:1017.

Ma, R. T. and P. B. Addis. 1973. *J. Food Sci.*, 38:995.

Maruyama, K., N. Kanemaki, W. Potts, and J. D. May. 1993. *Growth Develop. Aging*, 57:31.

Nakamura, R., S. Sekoguchi, and Y. Sato. 1975. *Poultry Sci.*, 54:1604.

Ngoka, D. A., G. W. Froning, S. R. Lowry, and A. S. Babji. 1982. *Poult. Sci.*, 61:1996.

Noll, S. L., M. E. el Halawani, P. E. Waibel, P. Redig, and K. Janni. 1991. *Poult. Sci.*, 70:923.

Norris, K. H. 1984. Reflectance spectroscopy, in *Modern Methods of Food Analysis*, K. K. Stewark and J. R. Whitaker, eds., Westport, CT: AVI Publishing Company.

Ortmann, R. 1975. *Anat. Embryol.*, 148:109.

Palombo, R., P. S. van Roon, A. Prins, P. A. Koolmees, and B. Krol. 1994. *Meat Sci.*, 38:453.

Pearson, A. M. and F. W. Tauber. 1984. *Processed Meats.* Westport, CT: AVI Publishing Co., p. 88.

Pearson, A. M., G. Harrington, R. G. West, and M. E. Spooner. 1962. *J. Food Sci.*, 27:177.

Purslow, P. P. 1985. *Meat Sci.*, 12:39.

Raj, A. B. M. 1994. *Br. Poult. Sci.*, 35:77.

Satterlee, L. D., G. W. Froning, and D. M. Janky. 1971. *J. Food Sci.*, 36:979.

Schaller, D. R. and W. D. Powrie. 1972. *Can. Inst. Food Sci. Technol. J.*, 5:184.

Schirmer, R. E. and A. G. Gargus. 1986. *Amer. Laborat.*, 18:30.

Schmidt, J. G. and F. C. Parrish. 1971. *J. Food Sci.*, 36:110.

Sosnicki, A. A. and B. W. Wilson. 1991. *Food Struct.*, 10:317.

Stiles, P. G. and A. W. Brant. 1971. *Poult. Sci.*, 50:392.

Swatland, H. J. 1975. *J. Anim. Sci.*, 41:78.

Swatland, H. J. 1982. *J. Food Sci.*, 47:1940.

Swatland, H. J. 1983. *Poultry Sci.*, 62:957.

Swatland, H. J. 1985. *J. Food Sci.*, 50:30.

Swatland, H. J. 1989a. *Internat. J. Food Sci. Technol.*, 24:503.

Swatland, H. J. 1989b. *Br. Poult. Sci.*, 30:787.

Swatland, H. J. 1990a. *J. Food Sci.*, 55:305.

Swatland, H. J. 1990b. *Can. Inst. Food Sci. Technol. J.*, 23:239.

Swatland, H. J. and S. Barbut. 1990. *Internat. J. Food Sci. Technol.*, 25:519.

Swatland, H. J. and S. Barbut. 1991. *Internat. J. Food Sci. Technol.*, 26:373.

Swatland, H. J. and S. Barbut. 1995. *Food Res. Internat.*, 28:227.

Tappel, A. L. 1957. *Food Res.*, 22:404.

Tarladgis, B. G. 1962a. *J. Sci. Food Agric.*, 13:481.

Tarladgis, B. G. 1962b. *J. Sci. Food Agric.*, 13:485.

van Hoof, J. 1979. *Vet. Quart.*, 1:29.

Vanderstoep, J. and J. F. Richards. 1974. *Can. Inst. Food Sci. Technol. J.*, 7:120.

Wardlow, F. B., L. H. McCaskill, and J. C. Acton 1973. *J. Food Sci.*, 38:421.

Webb, N. B., F. J. Ivey, H. B. Craig, V. A. Jones, and R. J. Monroe. 1970. *J. Food Sci.*, 35:501.

Weidemann, J. F., G. Kaess, and L. D. Carruthers. 1967. *J. Food Sci.*, 32:7.

Whiting, R. C. 1988. *Food Technol.*, 42:104.

Wolman, M. and F. H. Kasten. 1986. *Histochemistry*, 85:41.

Yannas, I. V. 1972. *J. Macromol. Sci. Rev. Macromol. Chem.*, C7:49.

CHAPTER 13

Improving and Sorting Meat Products

THE R&D ENVIRONMENT

THE MEAT INDUSTRY handles an astronomical tonnage of product on a global scale, and sales of meat in most developed countries are important items in the national economy. But profit margins often are very tight—at least, that is what they tell us. A high-volume industry with a low profit margin is not likely to spend much on R&D (research and development). Another gloomy fact is that the meat industry is a traditional industry, unable to bring a brand new product to a world of eager customers, as did the inventors of personal computers, compact disk players, and a host of other inventions that made overnight millionaires. Thus, while high-tech industries may plough back a high proportion of their returns into R&D, the meat industry tends to rely on its equipment suppliers to do this on their behalf. Equipment manufacturers try to keep up to date technologically and scientifically, and sometimes they take risks and innovate. Meat industry professionals, for their part, tend to shop around very carefully, learning through the grapevine what their competitors are doing and then making a safe, major purchase that can be justified to company financial officers. Thus, we find similar equipment being used in a similar way in major plants around the globe.

There are, however, some national variations on this theme. The meat industries in countries with both a large domestic product and a large domestic market, the United States being the obvious example, tend to act conservatively. Sometimes, packing plants are a security investment for a multinational company that also controls some high-risk companies, and it would be an unwise chief executive officer who got the two confused. Other packing plants are the remains of once-mighty conglomerates, sold at a bargain price, stripped down, and run leanly and efficiently. R&D staff are the first to be fired as the plant adjusts to its new lean image. Neither the plant kept as collateral for more risky ventures nor the lean and efficient remainder plant are likely to spend much on R&D without a pressing reason. If a multinational does indulge in R&D, usually, it is kept close to the home base, leaving foreign subsidiaries bereft of R&D capacity.

At the other end of the range in R&D roulette are the plants in countries desperate for export dollars. Countries lacking mineral wealth, but well-endowed with land or a tradition for agriculture, may elevate their meat industries to a high level of strategic importance. Coupled with vertical integration from producer to exporter or with a cooperative farming and marketing system, this may produce a very different frame of mind, where research is applauded, instead of ignored, and where funding for a sound idea may be regarded as security for the future. R&D at this end of the spectrum is doubly powerful, because those who commission and fund it are highly likely to use it.

So how does on-line evaluation of meat fit into this complex global industry? There are several connections, foremost of which is the quality revolution: the one in which customers finally revolted against poor design, obsolescence, and shoddy construction in their new cars and appliances. Foreign competition shook up the industries of complacent nations and brought us all into the era of product quality optimization, said by some to be the greatest advancement in manufacturing since the automation of assembly lines. According to this wisdom, to compete effectively, all aspects of design, method of construction, and choice of materials must be subordinated to producing the highest possible quality at the lowest possible price, combined with paring away stockpiles and replacing them with just-in-time production methods. There is no doubt that this quality revolution has transformed competitive manufacturing industries around the world, finally reaching the meat industry under the guise of value-based marketing (Cross and Savell, 1994). A consensus reached by the U.S. meat industry is that the industry should pursue research and development of an instrument for the assessment of carcass value (lean yield, marbling, and maturity).

HACCP (Hazard Analysis Critical Control Points; Sofos, 1993) has become essential in anything connected with the microbiological safety of meat, and, with newly trained and more knowledgeable personnel, now the same concepts are easier to apply to other areas, such as the control of meat quality. Our basic problem in attempting this, however, is the difficulty of predicting the intrinsic yield and quality of meat from a subjective judgement of the live animal or the external appearance of its carcass. Thus, the on-line evaluation of meat attempts to provide this information, and on-line evaluation is poised to change the way we do business in the meat industry.

Another major stimulus to the adoption of on-line evaluation is the gradual spread of vertical integration in the meat industry. When the feed suppliers and animal breeders become linked economically with meat producers and processors and when all are linked to their own retail outlets, there is a strong incentive to look critically at the final product and how it is judged by customers. On-line evaluation of meat attempts to provide information to enhance the competitiveness of vertically integrated systems. Once customers have a choice, producers may no longer have one: either producers take meat quality seriously, or they fall by the wayside. Quite likely, however, a sermon is being preached to the converted, because anyone reading this book already will have seen the light, but it does no harm to state our position clearly.

INFORMATION FLOW

Animal and Meat Tracking

Tracking individual animals from conception, through the production system all the way to the meat counter, is a costly but wise investment. Once the system is in place, animal production and meat marketing may be integrated and programmed to reach competitive objectives using the flow of information from on-line meat evaluation. Bjerklie's (1994) description of the Origen system developed by Coleman Natural Meats of Denver, Colorado, shows how it may be done with bar codes.

Every animal is stapled with an eartag showing a barcoded animal tracking number (ATN) shortly after birth. The tag is scanned during feeding to record days on feed, market weight, and feed sources. At slaughter, the ear tag is used to generate four new ATN tags, one for each quarter of the carcass, later proliferating to twenty-two tags that identify all the major regions of the carcass with the ATN, weight, time of entry into the plant, and USDA quality grade and, finally, generating as many box labels as are required to identify the meat within. Thus, a batch of ground beef carries all the ATN codes necessary to identify the sources of the batch.

The system would not be feasible without laser scanned bar codes, which, at present, are scanned manually. This accounts for much of the cost and makes large-scale implementation difficult, but the advantage to smaller plants of tracking systems cannot be ignored. Large plants have achieved dominance by economies of scale, but, if they fall behind in quality control technology, the competitive formula could change. To the bystander on the sidelines of the industry, it is rather like watching the role of the battleship in naval history, achieving prominence by an economy of scale and then losing it to smaller, more controllable types of ship.

Another approach to tracking is the use of radio frequency identification (RFID), which, although considerably more expensive, has many advantages (Stewart, 1993). RFID transponders may be attached to an ear or collar or implanted subcutaneously and may be read from any position within 0.5 m. The transponder contains a tuning circuit (inductor and capacitor) responding to and obtaining power from excitation signals at 120 to 135 kHz. This enables an integrated circuit to transmit encoded binary information. The main advantage of RFID during production is that individual animal feed intake may be monitored automatically; the disadvantage is at the abattoir. Even though the transponder may be correlated with another transponder to carry the animal identification on the gambrel, the system must be switched to bar codes to allow complete product identification.

Incorporating Information on Meat Quality

On-line meat evaluation may have an immediate application when selecting some of the total line product for a value-added treatment such as preparing dry, cured ham (Garrido et al., 1994), but it is very difficult to use information on meat quality for the total line product unless there is a system in place for tracking animals, carcasses, primal cuts, and products. But tracking systems are now in place in many modern abattoirs. At present, this information flow is dedicated mainly to information on the quantity of meat, enhancing inventory control, shipping, and billing. Adding information from on-line evaluation of meat could enable a well-organized plant to compete in a new dimension— meat quality.

For example, the only beef that is reliably tender at the present time is handpicked, aged under optimal conditions, and sold to premium customers or expensive restaurants. Making a comparable product available to ordinary customers at a modest premium could do something to restore per capita beef consumption to its former levels. Arguing against this are the experiences of those who may have tried to establish a premium product without the benefit of an effective method of on-line meat evaluation. Four things go wrong. First, while the incidence of tough beef may decline in a premium brand, it cannot be reduced to near zero without an effective method for imposing proper quality control procedures. Second, the extra cost to the customer may be more than the real increase in value of the product. Third, disillusioned customers question why all beef is so unreliable, even the premium product. Fourth, since an ignorant cook can ruin even the best beef on the planet, customer education must be included in the overall marketing package. On-line evaluation of meat ameliorates the first three of these problems and makes the fourth problem our first priority, which, rightly, it should be.

Feedback of meat-yield information to the producer is important to discourage the production of carcasses with a low yield or an awkward size that cannot easily be sold. In many countries, the producer is rewarded financially for producing an optimum carcass in terms of fatness and size, but seldom is any attention given to meat quality. Seldom are tough beef and PSE pork carcasses downgraded at the abattoir, provided they look alright from the outside. Someone else farther down the chain bears the loss — often the customer.

Look-up Tables

Figure 13.1 introduces the look-up table, found at the heart of many meat grading systems. As we saw in Chapter 1, there is not much in a carcass except muscle, fat, and bone. The proportion of bone is fairly constant, so if the proportion of fat is high, then the proportion of muscle must be low and vice versa. This enables the proportion of muscle to be

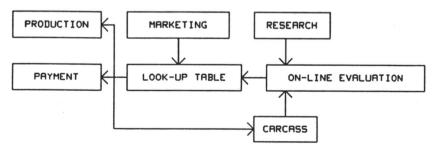

Figure 13.1 *Block diagram showing the importance of the look-up table software in coordinating animal production, payment of producers, research, and improvement of on-line meat evaluation under the direction of marketing.*

estimated from a combination of carcass weight and back-fat thickness, as considered in Chapter 1. In the future, we may expect to see continued improvements in the accuracy of estimating the muscle mass on a carcass as the sample base for the fat-depth measurement is increased, as well as incorporating direct assessments of the muscle.

Most aspects of meat quantity and quality are continuous variables, such as fat depth and meat tenderness. Meat grades, however, are discontinuous variables, a series of step functions. There are two ways for a computer to produce a discontinuous grade from a mixture of continuous variables from the on-line evaluation of meat, either by arithmetic calculation or by looking up the answer in a table. If the relationship between the on-line input and the grade output is complex and if there are only a few possible grades, as is usually the case, then the look-up table is the best option. The grade derived from the look-up table may be fed back to the producer as a payment for the lean meat produced or as information for breeding or planning the best marketing strategy. The grade also may be marked on the carcass for the benefit of retailers and customers (feedforward).

Few things remain constant for long in the world of commerce, and, when market conditions change and different grades become necessary, all that is required is a change in the look-up table, not a change in the on-line grading hardware. Separating the on-line grading hardware from the look-up table facilitates changing the look-up table to issue new grades, but it also enables the on-line grading hardware to be changed while keeping the same look-up table. Many different on-line technologies have been described in this book, some already operating successfully and others wildly impractical at present. But nearly all ultimately significant inventions are impractical at first, and it is difficult to believe that technology is not going to keep evolving in a major industry like ours. Thus, it is important not to remain fixed on any one type of on-line hardware. New systems, perhaps less expensive, more accurate, or providing extra information, will eventually become available. When they do, it will be relatively simple to assimilate them if grading systems are based on look-up tables. Future look-up tables might be multidimensional, with yield in the first two dimensions and quality in extra dimensions.

Continuous versus Batch Processing

On-line evaluation of meat may have a greater impact in continuous processing than in batch processing. With batch processing, ingredients for meat products are mixed using least-cost formulation to decide the relative proportions and then launched into the processing line. If there is an error, then the whole batch is affected, and the loss is a function of the magnitude of the error and the mass of the batch, which may be quite considerable. Essentially, everyone hopes for the best once the batch is mixed.

With continuous processing, on the other hand, mixing is regulated in real time so that the ingredient mixture is controlled, consistent, and cost-effective. Augers of high-priced meat ingredients with high-protein functionality are balanced against augers of low-cost filler. Instead of a master butcher's magic thumb testing a minuscule fraction of a batch, all of the continuous stream of ingredients, as a slurry or emulsion, passes the on-line sensors that control both mixing and later processing operations. In the relatively simple system shown in Figure 13.2, auger 1 contains the meat with a high-protein functionality, while auger 2 contains filler, and the two streams meet in the mixer. Auger rates are controlled by feedback to keep the protein functionality of the ingredient mixture at a controlled level, ensuring a reliable emulsifying capacity, water-holding capacity, color, low rancidity potential, fat content, filler content, or cookability of the future product.

As well as a feedback of information to the augers in Figure 13.2, there is also a feedforward to a processor, such as an oven or smokehouse, and to the packaging equipment. Feedback control is subject to certain limitations. If the protein functionality of the meat in auger 1 is inadequate because of the cost or unavailability of a superior ingredient, product quality will start to deteriorate, not erratically in large batches, but slowly and continuously in a way that can be predicted from the on-line sensor. Thus, the feedforward of information may be remedial

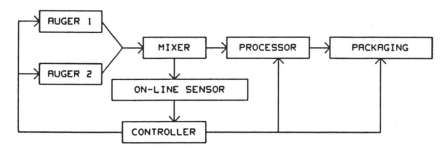

Figure 13.2 *Block diagram showing feedback and feedforward information flow in a continuous processing line.*

in nature, perhaps cooking the product for longer at a lower temperature, or it may be protective, perhaps changing the packaging or product brand label.

Computer Integrated Manufacturing

Computer integrated manufacturing (CIM) is a system for the integration of all aspects of manufacturing, especially planning, administration, and process control. It provides an overall environment in which on-line evaluation of meat could flourish, identifying the things that need to be measured and justifying the research to find how to measure them. Unless something can be measured, it cannot be controlled or improved, but technology transfer is far easier if the application is known before the solution is discovered or invented.

Many progressive meat plants employ computer-based control systems for isolated operations, but few have attempted complete integration. One of the best documented is the Danish Tulip–Vejle Nord project commissioned in 1989 and producing 200 to 250 tons per day of sausages, luncheon meat, chopped ham, and liver paste. The plant has six subsystems:

(1) Raw material store: The raw meat ingredients are stored in trays, each quantified for mass, fat content, and processing characteristics. The trays are stored in vertical racks, which then are retrieved and moved on-line when required. Thus, substitution of equivalent materials may be used to maintain a least-cost formulation with a consistent product quality.

(2) Comminution: Separate lines for canned products and sausages are filled with batches of materials that are comminuted and mixed, with the automatic addition of meat, water, starch, additives, and blood by-products.

(3) Sausage cooking and smoking: Conveyer systems are used to move products through continuous processing lines, with separate lines for artificial and natural casings.

(4) Canning: This operation receives product directly from the comminution subsystem, converging on a continuous supply of cans and lids for completely automated filling, closing, and washing.

(5) Cooking: Automatic horizontal retorts are operated in batch mode.

(6) Packing: Labeling, packing, boxing, and pallet packing lines operate as parallel continuous lines.

The CIM concept is operated as a multilevel, closed-loop control system. The top-down control hierarchy is divided into four levels:

(*1*) Management of product orders of the whole enterprise

(*2*) Production plans for the factory

(*3*) Job plans for each product production system

(*4*) Process control at each work station

The bottom-up production hierarchy has a matching structure:

(*4*) Process monitoring

(*3*) Job or batch status

(*2*) Production status

(*1*) Resource utilization

The control loops operate laterally at each of the four levels, and vertically through the control hierarchy. Thus, in this system, the role of on-line meat evaluation may be localized to process control and resource utilization (evaluation of raw materials).

Technology Transfer

All of us involved in the meat industry need to work energetically to improve the transfer of technology from the research laboratory out into the real world of industry. Industry professionals have the legitimate excuse that their lives are dominated by immediate problems, often amounting to the routine management of crises, leaving little time to wrestle with the conflicting nuances of meat science. Like any other science, meat science is always evolving: in practical terms, there is a sustained conflict between old and new ideas, one team of expert witnesses slugging it out against another team. The best theory often takes years to decide, by which time the audience of industry personnel has departed long ago in a state of confusion. Argument is the engine of progress and should not be suppressed, but meat scientists would win a lot more support from their colleagues in industry if some of their scientific battles could take place in the real world, as well as in the laboratory. This is something we can achieve with the on-line evaluation of meat.

From the laboratory perspective, we can develop on-line methods of meat evaluation and substantiate their validity using reliable laboratory methods. Then we can install the technology on-line, collecting large amounts of data that encompass the great biological ranges in meat yield and quality. Thus, we could use large commercial plants as our laboratories for large-scale experiments that deal with the immense complexities of routine operations. This could provide vital information to enhance commercial competitiveness, making it far more attractive for industry personnel to become involved in the process of scientific debate.

On-line evaluation of PSE provides an excellent method of in-house research to answer questions relating to transport and slaughter methods, as well as enabling the comparison of different sources of animals. Correcting all the obvious abuses that create PSE carcasses from normal pigs is costly, but an on-line experiment may show the advantage attainable, thus enabling the cost-benefit ratio to be established. Cost savings is another angle: perhaps some of the costly processes we take for granted are not really contributing much to the yield or quality of the product. Electrical stimulation of beef to guard against cold-shortening certainly may be cost-effective in certain situations, but can it be proven for a specific installation?

In-house Research on PSE

One gains a strong impression that many meat industry professionals now have a good understanding of local causes of PSE pork and ways to avoid it. Some years ago, pigs were severely stressed during marketing from the farm to the abattoir, were herded with electrical prods under conditions that favored aggressive activity, forced into dark pens and up slippery steep ramps, and stunned ineffectively. Then the carcasses were taken lazily through scalding tanks, eventually leaving the kill floor to be tightly packed in relatively warm meat coolers. These abuses now have been corrected in competitive abattoirs. Paradoxically, however, these improvements did not always result in dramatic reductions in PSE, largely because they often took many years to implement, during which time the pig population and production methods kept evolving. In the realm of opinion, rather than provable fact, it seems likely that intensive production systems are more likely to produce PSE pork than a commercially less competitive production system. Free-range pigs used to exercise and temperature fluctuations and brought to market as a social group may produce less PSE than pigs reared intensively (Warris et al., 1983; Barton-Gade and Blaabjerg, 1989), although this cannot always be detected (Van Der Wal et al., 1993; Jones, 1993).

Although it may be possible to charge a premium for beef guaranteed to be tender, this may not necessarily be true for premium pork guaranteed free from PSE. In Denmark, the MQM probe was used to select pork with a good water-holding capacity and well-developed marbling, and the meat was aged for seventy-two hours before sale. Despite extensive brandname advertising, the 15% price premium was more than customers were prepared to pay, and the product was discontinued (Andersen et al., 1993). However, Denmark has a relatively small domestic market with a high degree of standardized cooperative meat production. In the much larger U.S. domestic market, with aggressive competition in both production and retailing, the outcome might be different. The critical factor is the magnitude of the quality difference between the normal and the premium product.

On-line evaluation of PSE provides an ideal way to test some of the improvements we have made in animal handling and slaughtering. In Figure 13.3, the solid lines show the fiber-optic reflectance of pork from pigs electrically prodded before slaughter and with simulated maximum reflex activity (Swatland, 1988). There was not much difference at forty-five minutes, but, by twenty-four hours postmortem, PSE was significantly ($P < 0.005$) worse in the prodded group with accelerated glycolysis. From data such as those in Figure 13.3, it is possible to predict the customer perception of meat and the predicted fluid losses

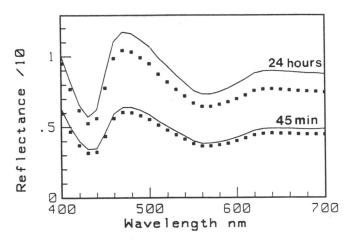

Figure 13.3 *Effect of electrical prodding and simulated reflex muscle activity on the development of PSE. Animals with accelerated glycolysis (solid lines) are compared with controls (solid squares) at forty-five minutes and twenty-four hours postmortem.*

during refrigeration and meat cutting (Chapter 8), thus quantifying the economic losses caused by prodding and poor slaughter technique. Converting data from one form to another to make predictions is a powerful tool in scientific research, but going one step further, converting scientific units and ratios into dollars and cents, is a powerful incentive for technology transfer.

Improving the Animals

The classical dilemma in attempting to select breeding stock to improve meat yield or quality is that it is very difficult to evaluate an animal while it is still alive and capable of breeding. Almost every angle on this problem has been exploited, from progeny testing to genetic markers. Ultrasonic evaluation of fat depth and muscle areas is widely used already. What is needed now is new technology. Much of the funding for this objective already has gone towards systems such as computer-assisted tomography and nuclear magnetic resonance imaging, but, unless there are unimaginable cost reductions and simplifications of these technologies, they are most unlikely to achieve any practical application in the mainstream of agriculture. Thus, just as with on-line evaluation of meat, the challenge is to achieve the objective on a low budget so that it has a hope of being applied in the real world. One possibility is the miniaturization of many of the techniques considered in this book so that they may be used in hypodermic needles. Bioelectrical impedance using needle electrodes has been tested satisfactorily for live pigs (Swantek et al., 1992), and it might also be possible to use optical probes.

It may be possible to measure connective tissues in living muscles via a hypodermic needle small enough for use on live animals, thus allowing genetic selection for muscle collagen content. The optical components shown in Figure 3.12 were connected to a hypodermic needle probe, and, with the optical fiber cut at the same angle as the tip of the hypodermic needle, signals from test charts were predicted from the long diameter of the elliptical optical window, exactly as outlined in Chapter 3. When a plastic optical fiber was linked with fluid to the test chart, the overall signal was reduced, but the pattern was unchanged. A quartz fiber did not show such a large drop in the signal. Signal peaks were deformed when obtained from meat samples, probably because of the elasticity and tensile strength of the meat microstructure, and signal peaks from septa of connective tissue were greatly increased in way-in transects, relative to way-out transects. Dynamic analysis of data from the depth sensor of the probe showed that this effect was caused by the delay before the needle burst through the stretched connective tissue. The design of a miniature optical probe mounted in a hypodermic needle, therefore, is a compromise between a sharp angle with easy penetration versus a blunt angle with high resolution. Thus, in the future, we may be able to look at collagen concentration and cross-linking in the live animal, eliminating those breeding stocks destined to produce offspring with tough meat.

Technologies of the Future

It is impossible to predict the future of any of the technologies reviewed in this book. Surely, progress will continue, but it is the cost-effective formula that is most difficult to predict. Devices such as optical fat-depth probes might conceivably disappear in the future, replaced by TOBEC or ultrasonic systems. On the other hand, their relatively low cost and decentralized distribution may give them a competitive advantage for many years to come, especially in small abattoirs. Nuclear magnetic resonance has been omitted almost completely from this book because it is too expensive and difficult to be considered as a routine on-line method for meat evaluation; however, there is no predicting what the future may hold, and this technology might one day become cost-effective. Chemical sensors are starting to appear, such as a glucose sensor for spoiled meat (Kress-Rogers et al., 1993), and these might evolve into on-line sensors for meat flavor and aroma. Thus, the only thing we can be sure of is that the whole discipline of on-line meat evaluation will continue to develop as commercial competition on the basis of meat quality is superimposed onto the existing economic framework of competition based on growth rate and feed efficiency.

THE BOTTOM LINE—CONSUMERS VERSUS CUSTOMERS

To understand the potential importance of on-line meat evaluation, it is necessary to understand the difference between consumers and customers. Consumers are an anonymous mass of people who are surveyed, taxed, and manipulated by advertising. Consumers will eventually buy all the shoddy goods and low-quality meat that are dumped on them. Customers, on the other hand, are individual people with a disposable income, who learn quickly what is and what is not worth buying. When displeased, they take their business elsewhere. Consumers who have a choice are elevated to the status of customers.

On-line meat evaluation enables us to integrate laboratory science with industrial technology in the single-minded pursuit of how to measure meat yield and quality — rapidly, reliably, and realistically, without damaging the meat that is measured. On-line meat evaluation enables all the well-established principles of quality control procedures to be applied to meat. The long-term objectives are to reward the producers of superior meat animals, to guarantee the quality of premium meat products, and to define ways of improving the overall quality of meat for our customers.

REFERENCES

Andersen, J. R., C. Borggaard, T. Nielsen, and P. A. Barton-Gade. 1993. *39th International Congress of Meat Science and Technology.* Calgary, Alberta, pp. 153 – 164.

Barton-Gade, P. and L. O. Blaabjerg. 1989. *Proceedings of the 35th International Congress of Meat Science and Technology.* Copenhagen, pp. 1002 – 1005.

Bjerklie, S. 1994. *Meat & Poultry, 40(7):26.*

Cross, H. R. and J. W. Savell. 1994. *Meat Sci., 36:19.*

Garrido, M. D., S. Banon, J. Pedauye, and J. Laencina. 1994. *Meat Sci., 37:421.*

Jones, S. D. M. 1993. *Meat Focus Internat., 2:396.*

Kress-Rogers, E., E. J. d'Costa, J. E. Sollars, P. A. Gibbs, and A. P. F. Turner. 1993. *Food Control, 4:149.*

Sofos, J. N. 1993. *Meat Focus Internat., 2:217.*

Stewart, R. C. 1993. *Meat Focus Internat., 2:415.*

Swantek, P. M., J. D. Crenshaw, M. J. Marchello, and H. C. Lukaski. 1992. *J. Anim. Sci., 70:169.*

Swatland, H. J. 1988. *Can. Inst. Food Sci. Technol. J., 21:494.*

Van Der Wal, P. G., G. Mateman, A. W. de Vries, G. M. A. Vonder, F. J. M. Smulders, G. H. Geesink, and B. Engel. 1993. *Meat Sci., 34:27.*

Warris, P. D., S. C. Kestin, and J. M. Robinson. 1983. *Meat Sci., 9:271.*

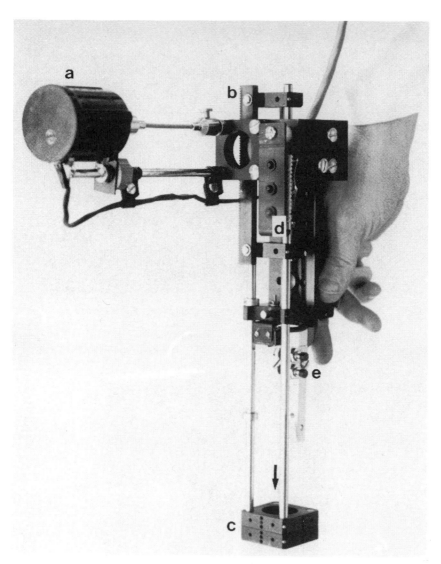

Plate 1 *Apparatus for the on-line evaluation of meat comes in all manner of shapes and sizes: this is a handheld prototype easily adaptable to any application in fiber optics. The round cylinder (a) is a precision potentiometer connected via a flexible shaft to a rack and pinion gear (b). As the probe is pushed down into the meat, the plate at the bottom (c) is pushed up, turning the potentiometer to give the depth in the meat of whatever has penetrated the meat through the hole in the center of the plate. The optics have been removed to show the electromechanical components more clearly, but they may be mounted anywhere in the region shown by the arrow. A set of three LEDs enables communication with the main controller (d). The operator's index finger is on a trigger switch, but there are two extra switches (e) mounted on the shaft that supports the optics (photo courtesy of Trina Koster).*

Plate 2 *At the other end of the range in size and complexity to the apparatus in Plate 1 is a major commercial installation for on-line evaluation of meat, the Danish Carcass Classification Center, which uses nine probes to maximize the information from each carcass (courtesy of Dr. J. R. Andersen and the Danish Meat Research Institute).*

Plate 3 *A closer look at the Danish Carcass Classification Center (courtesy of Dr. J. R. Andersen and the Danish Meat Research Institute).*

Plate 4 *A pork carcass held against a robotic probe, which moves quickly down the carcass to the measuring position located ultrasonically relative to the skeleton (robot courtesy of Dr. A. Goldenberg; operator Norm Auliffe; photo courtesy of Office of Research, University of Guelph).*

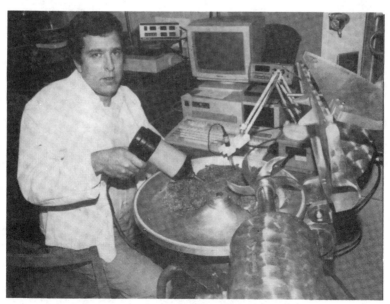

Plate 5 *Using a Colormet fiber-optic probe to monitor optical changes during meat processing operations (operator Dr. S. Barbut; photo courtesy of Office of Research, University of Guelph).*

333

Plate 6 *Using a Colormet fiber-optic probe to measure meat color in an intact carcass (operator Dr. M. Irie; photo courtesy of Office of Research, University of Guelph).*

Plate 7 *Using an optical window Colormet to quantify cyanosis in the veterinary inspection of poultry (operator Dr. J-P. Vaillancourt; photo courtesy of Office of Research, University of Guelph).*

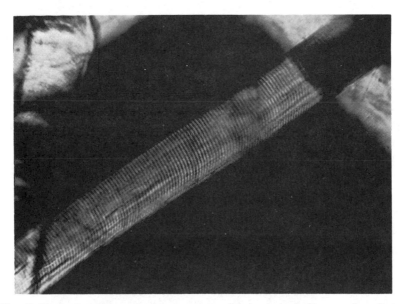

Plate 8 *The basic cellular unit of meat: a striated skeletal muscle fiber seen by polarized light microscopy, showing bright A band striations and dark I bands.*

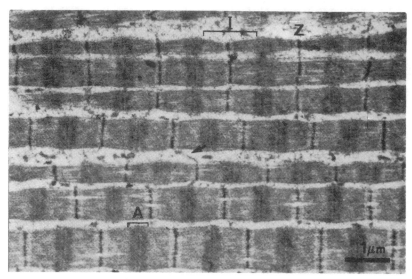

Plate 9 *Within a muscle fiber are the myofibrils. At much higher magnification than Plate 8, the electron microscope shows that A and I bands are determined by overlapping myofilaments. At the midlength of the I band is the Z line, which holds thin myofilaments in place. Between myofibrils (arrow), the cytoskeleton normally keeps the Z lines neatly lined up, but the alignment has been lost from this strip of pork being tested in a rigorometer.*

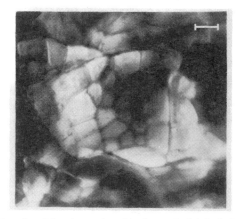

Plate 10 *A thin slice of pork has been placed in front of a bright light to show the muscle fibers (bar scale = 0.1 mm). Fibers with a high myoglobin content appear darker than those with low myoglobin, and a diagnostic feature of pork versus other meats is evident: muscle fibers in the centers of their bundle have more myoglobin than fibers around the outside.*

336

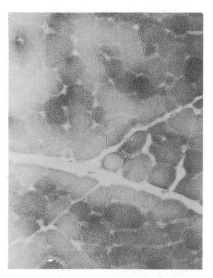

Plate 11 *A histological transverse section of beef has been reacted histochemically to show the distribution of myoglobin, which differs slightly between fibers.*

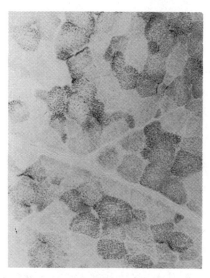

Plate 12 *This is a serial section to that in Plate 11, allowing the same muscle fibers to be located. Fibers have been reacted histochemically to show the distribution of mitochondria by the succinate dehydrogenase reaction. Fibers with a high myoglobin content also tend to have a high mitochondrial content, revealing their differential metabolism in relation to a range of commercially important properties such as color, taste, and rancidity after processing.*

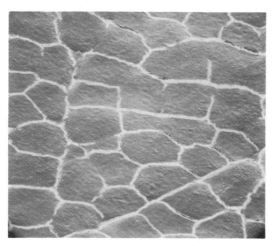

Plate 13 *The distribution of fluid in meat is highly variable, leading to serious economic losses from PSE meat. In the live animal, normally, there is very little fluid space between individual muscle fibers, as shown here in a scanning electron micrograph made transversely across the muscle fibers. Some of the fibers are pushed so tightly together that the gap between them cannot be seen.*

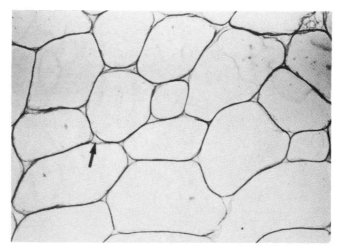

Plate 14 *Fluid-filled spaces between the muscle fibers appear as muscle develops rigor mortis and is converted to meat. In this transverse section of pork muscle immediately after slaughter, the contents of the muscle fibers have been removed to leave only the cell membranes, which show as black lines. The arrow indicates the position of fluid exudate, just starting to accumulate in the triangular spaces between the compressed muscle fibers.*

338

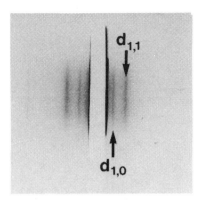

Plate 15 *Much of the fluid lost from meat originates from between the myofilaments of the myofibrils within the muscle fibers. The lateral spacing of the myofilaments may be measured most accurately by X-ray diffraction, the results of which then may be used to validate on-line probes. This is an X-ray diffraction pattern from pork, showing the $d_{1,0}$ and $d_{1,1}$ spacing (courtesy of Dr. B. M. Millman).*

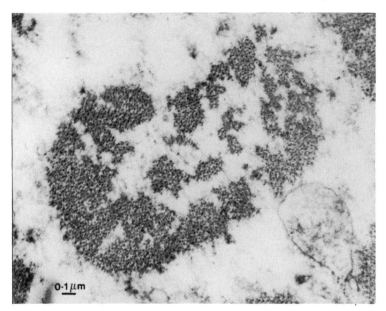

Plate 16 *Eventually, when the pH gets very low and myofilaments come very close together, there may be a massive fluid loss from within the myofibril, as seen here in this electron micrograph of a transverse section of pork. Fluid loss in this exceptional case has been so severe that the myofibril has fragmented internally.*

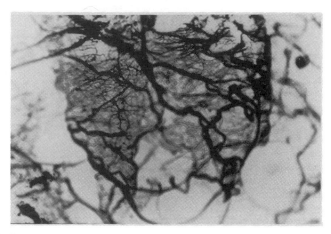

Plate 17 Connective tissue fibers are a well known cause of meat toughness. The small reticular or Type III collagen fibers, seen here blackened with silver, are part of the endomysium on the surface of a muscle fiber.

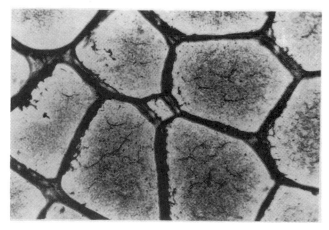

Plate 18 When meat is sectioned transversely, the connective tissue fibers on the surface of muscle fibers appear to form an endomysial tube around the muscle fiber.

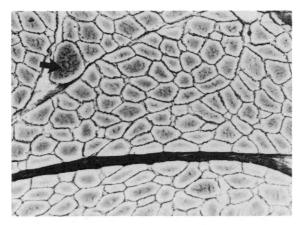

Plate 19 At lower magnification than in Plate 18, numerous individual muscle fibers surrounded by their endomysial tubes may be seen. Running horizontally across the lower field is a seam of perimysial connective tissue, which could have been detected on-line using UV fluorescence fiber optics. The arrow indicates a giant fiber.

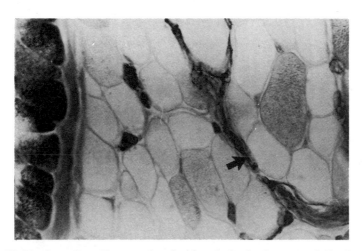

Plate 20 The formation of lactate after slaughter is the dominant metabolic change as muscles are converted to meat. Lactate is difficult to visualize, but the disappearance of glycogen that gives rise to the lactate may be seen with the periodic-acid Schiff reaction histochemically. There is a whole experiment in this one final plate. Down the left side are transversely sectioned muscle fibers, part of a bundle of fibers that have been inactive after slaughter and still retain most of their initial glycogen. Nearly all the fibers to the right have exhausted their glycogen, having converted it to lactate, and they appear empty. The reason why this right-side bundle of muscle fibers was active has been revealed by the fortuitous staining of nerve axons, indicated by the arrow. The nerve on the cut surface of the meat had been firing irregularly and had caused the fibers of this bundle to contract, thus accelerating their rate of glycolysis.